FET Circuits

Rufus P. Turner, Ph.D.

HOWARD W. SAMS & CO., INC.
THE BOBBS-MERRILL CO., INC.
INDIANAPOLIS · KANSAS CITY · NEW YORK

FIRST EDITION

SIXTH PRINTING—1973

International Standard Book Number: 0-672-20585-8
Library of Congress Catalog Card Number: 67-27005

Preface

The field-effect transistor is a comparatively new semiconductor device. Its chief merit is its high input impedance, a feature which makes it more adaptable than the conventional (or *bipolar*) transistor to tube-type circuits. But it also possesses those qualities that recommend *all* transistors: simplicity; small size; ruggedness; instant operation; high overall efficiency; and freedom from hum, microphonics, and, in most cases, generation of heat.

Now that the FET is commercially available at comfortable prices, designers and experimenters at all levels are showing interest in its applications. This book attempts to meet the demand for practical information on the subject.

The numerous tested circuits described on the following pages cover the field of applications from amplifiers to test instruments and will be especially interesting to experimenters and hobbyists. (A great many of the circuits contain only one or two FET's.) This collection is made up of circuits which survived impartial tests to eliminate those which gave only marginal performance.

The reader may use these circuits singly or in such combinations as his needs suggest. The present collection by no means exhausts the possibilities of FET application; therefore, many of these circuits might also serve as idea generators.

The author is grateful to Siliconix Incorporated (Sunnyvale, California) and to Texas Instruments, Inc. (Dallas, Texas), who kindly supplied operating data and generous samplings of their field-effect transistors for use in developing the circuits described in this book.

RUFUS P. TURNER

Contents

CHAPTER 1

FUNDAMENTALS OF THE FET 9
The Field Effect—Background of the FET—FET Structure and Operation—FET Performance—FET Ratings—FET Types

CHAPTER 2

AMPLIFIER CIRCUITS 19
Simple Small-Signal A-F Preamplifier—Single-Stage, Low-Gain A-F Preamplifier — Single-Stage, Medium-Gain A-F Preamplifier — A-F Source Follower—Two-Stage, R-C–Coupled A-F Amplifier—Two-Stage, Transformer-Coupled A-F Amplifier—High-Impedance Input For Bipolar Transistor—Conventional Phase Inverter—Paraphase Phase Inverter—Dual-Channel Audio Mixer—L-C–Tuned Bandpass A-F Amplifier—R-C–Tuned Bandpass A-F Amplifier—R-C–Tuned Band-Suppression (Notch) A-F Amplifier—Video Amplifier—455-kHz I-F Amplifier—Auxiliary Headphone Amplifier—Simple Audio AGC Amplifier—Simple D-C Amplifier

CHAPTER 3

OSCILLATOR CIRCUITS 43

Transformer-Feedback A-F Oscillator—L-C-Tuned A-F Oscillator—Phase-Shift A-F Oscillator—Single-Frequency Wien-Bridge A-F Oscillator—Drain-Coupled Multivibrator—Source-Coupled Multivibrator—Self-Excited R-F Oscillator—Conventional Crystal Oscillator—Pierce Crystal Oscillator—100-kHz Crystal Oscillator—Colpitts A-F Oscillator—Code-Practice Oscillator

CHAPTER 4

RECEIVER AND RECEIVER-ACCESSORY CIRCUITS 59

Superhet Broadcast Receiver—Regenerative Broadcast Receiver—All-Wave Regenerative Receiver—Single-Frequency, Crystal-Controlled Converter — Beat-Frequency Oscillator — Q-Multiplier — "Selectoject"—Audio Squelch Amplifier

CHAPTER 5

TRANSMITTER AND TRANSMITTER-ACCESSORY CIRCUITS 73

Single-Stage Crystal-Controlled Transmitter—Single-Stage 0.8-Watt Crystal-Controlled Transmitter — Two-Stage Transmitter — Push-Pull Final Amplifier—100-MHz Grounded-Gate Amplifier—Single-Ended Frequency Multiplier—Push-Push Doubler—Four-Band Exciter—Balanced Modulator—C-W Monitor—Modulating the FET Transmitters—Single-Control ("TNT") Transmitter—Phone Monitor

CHAPTER 6

CONTROL CIRCUITS 93

D-C Relay—A-C R-F Relay—Touch-Plate Relay—Amplified Photocell—Photo-FET Relay—Timer—FET as Voltage Variable Resistor—Constant-Current Source—Sensitive A-C A-F Relay—Sound-Operated Relay— Modulated-Light Detector/Amplifier—Phase Shifter

CHAPTER 7

INSTRUMENT CIRCUITS 111

Single-FET Electronic D-C Voltmeter—Balanced Electronic D-C Voltmeter—Electronic A-C Voltmeter—Electronic A-C Voltmeter/Millivoltmeter—Crystal Dual-Frequency Standard—100-kHz/10-kHz Secondary Frequency Standard—Direct-Reading Audio-Frequency Meter—4.5-MHz Sound-Marker Generator—Variable-Frequency Audio Oscillator—Dip Oscillator—A-F Milliammeter—Inductance Checker—Substitution-Type Capacitance Meter—Impedance Meter—Heterodyne Frequency Meter—Harmonic-Distortion Meter—A-F R-F Signal Tracer—Sensitive Light Meter—Geiger Counter

APPENDIX A

TERMINAL GUIDE FOR DIODES AND TRANSISTORS 151

APPENDIX B

MANUFACTURERS OF SEMICONDUCTORS 153

APPENDIX C

DIRECTORY OF MANUFACTURERS 155

INDEX 157

1

Fundamentals of the FET

The *field-effect transistor,* because it behaves very much like a vacuum tube, opens new approaches to semiconductor circuit design. Some circuits in which regular transistors either operate poorly or require special components now operate efficiently and without modification with field-effect transistors.

We can wish that the field-effect transistor had come first. Many transistor circuits then would have been much simpler than at present, and the transition from tubes to transistors would have been less painful than it was. The regular transistor, as a later development, would have had many special-purpose applications. Nevertheless, the late coming of the field-effect transistor detracts little from its usefulness.

Field-effect–transistor theory differs somewhat from that of the regular transistor. However, it can be presented in simple terms. Obviously, the circuit designer functions with increased ease and profit when he understands how this new semiconductor device works. This chapter presents the necessary elementary theory so that this understanding can be developed.

THE FIELD EFFECT

For more than a century electrical men occasionally pondered the idea of controlling current in a conductor by applying an electrostatic field across the latter. And undoubtedly a great many of them tried it—unsuccessfully. Indeed, sooner or later every experimental-minded schoolboy gets around to mounting a resistor or a length of insulated wire between the plates of a capacitor to see if a voltage applied to the plates will affect the current in the resistor or wire. But no practical device emerged from any early experiments of this kind.

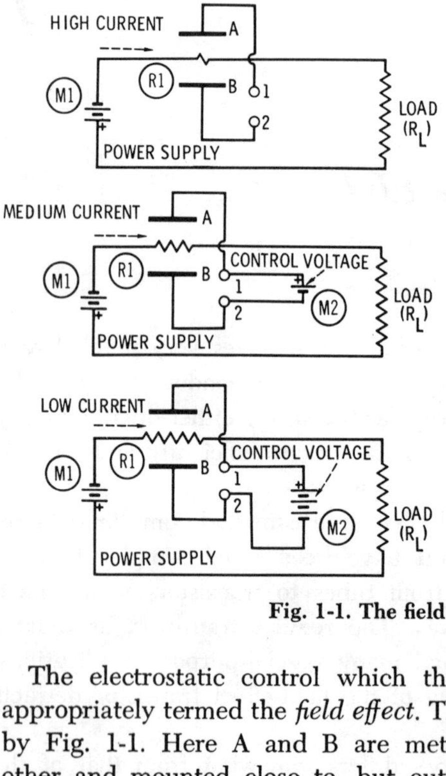

(A) Zero control voltage.

(B) Medium control voltage.

(C) High control voltage.

Fig. 1-1. The field effect.

The electrostatic control which these experimenters sought is appropriately termed the *field effect.* This phenomenon is illustrated by Fig. 1-1. Here A and B are metallic plates parallel to each other and mounted close to, but out of contact with, some sort of resistor (R1) which is sensitive to an electrostatic field. If a voltage is applied to the plates through terminals 1 and 2, the field set up between the plates penetrates the resistor. A power supply, represented by battery M1, passes current through the sensitive resistor (R1) and a load resistor (R_L) in series. When the control voltage is zero (Fig. 1-1A), no field exists between the plates, and resistance R1 is a very low value. Consequently, there is a high current through the load. When a finite control voltage subsequently is applied to the control terminals (1, 2), R1 will change and so will the load current. Thus, when the control voltage (battery M2) has some medium value (Fig. 1-1B), R1 assumes a somewhat higher value than before, and the load current decreases to a medium level. Similarly, when the control voltage is high (Fig. 1-1C), R1 becomes high and the load current is reduced to a low value. If the control voltage is very high, R1 becomes infinite and the load current falls to zero. In this way the load current is modulated by the control voltage. Another way of thinking about the field effect is to suppose that the control voltage *decreases* the sensitive resistance.

The advantage of such a field-effect device is its voltage-responsive character. That is, ideally no current is drawn from the control-voltage source, although the controlled current may be sizable. This is the same advantage offered by the vacuum tube. Unlike the tube, however, the field-effect device requires no heater. The field-effect transistor is the first practical example of such a device.

BACKGROUND OF THE FET

Early searchers for a crystal triode were frustrated in their attempts to use an electrostatic field to modulate current in a semiconductor. They had hoped in this way to imitate the vacuum tube, and their later studies seeking to find out why the arrangement refused to work led to the invention of the regular transistor. That this transistor became immensely useful is history. But its low input impedance resulted in (1) inability to replace the tube in some circuits, (2) need for tapped coils and step-down transformers, and (3) difficulty for many persons to shift from voltage-amplifier–tube thinking to current-amplifier–transistor thinking.

Despite the success of the regular transistor, work continued along the original line. Accordingly, several experimental field-type transistors appeared in the United States and in Europe during the 1950's. (Dr. William Shockley, one of the inventors of the first transistor, reported some of his work along this line in an article, "A Unipolar Field-Effect Transistor," in the November, 1952, issue of *Proceedings of the I. R. E.*) The modern field-effect transistor (FET) is the culmination of this continued investigation.

The FET has supplied the high input impedance, through electrostatic control, that was sought by the original researchers. Being more compatible with the vacuum tube than is the regular transistor, the FET gives promise of eventually replacing the tube—and the regular transistor—in many circuits used in electronic equipment.

FET STRUCTURE AND OPERATION

Fig. 1-2 shows the basic structure of the FET; and while this is not the actual configuration of some commercial models, it is accurate enough for explanatory purposes. The heart of the device is a thin bar or wafer of silicon (infrequently, germanium) with an ohmic (nonrectifying) contact (A, B) at each end. The end-to-end electrical path through this bar is termed the *channel.* If the silicon is N type, as in Fig. 1-2A, a P region is processed into each face of the bar in such a way that each such region is parallel to the other, and a connection (C, D) is made to each. These regions

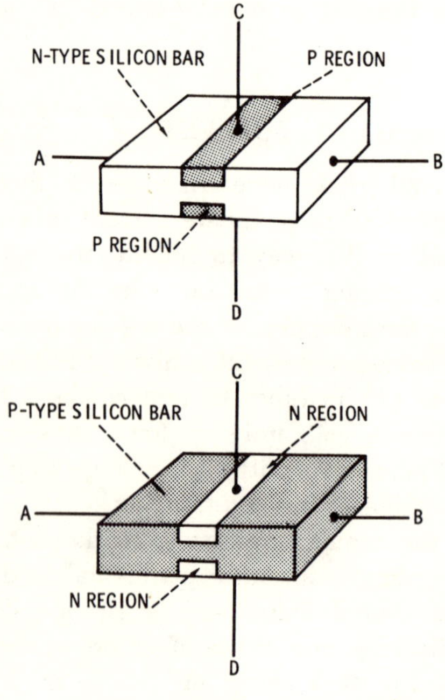

(A) N-channel.

(B) P-channel.

Fig. 1-2. Basic structure of a field-effect transistor.

are termed the *gates*. (In most commercial FET's, the gates, if two are used, are wired together internally and connected to a single terminal.) The completed transistor is given the name N-channel field-effect transistor (NFET). If, instead, the silicon is P type, as in Fig. 1-2B, then the gates are N regions and the transistor is named P-channel field-effect transistor (PFET).

When a d-c voltage is applied between A and B, the current carriers (electrons in the N-channel and holes in the P-channel FET) flowing through the bar must pass through the channel between the two gate electrodes. The anode terminal (A) is termed the *drain*, and the cathode terminal (B) the *source*. In a *symmetrical FET*, either terminal may be the source, and the other the drain. The drain is equivalent to the plate of a tube or the collector of a regular transistor; the source is equivalent to the cathode of a

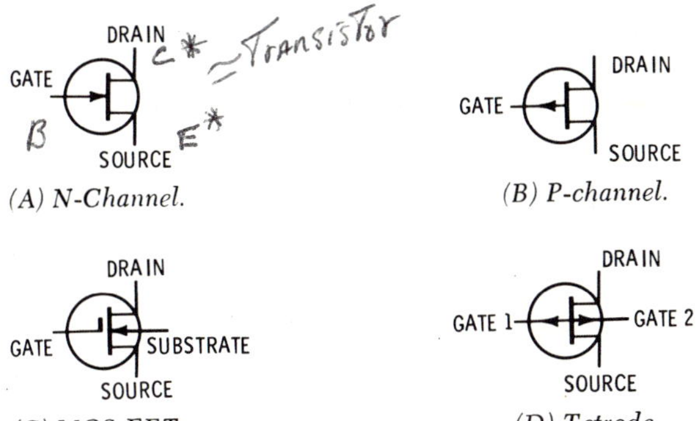

(A) N-Channel. *(B) P-channel.*

(C) MOS FET. *(D) Tetrode.*

Fig. 1-3. FET symbols.

tube or the emitter of a regular transistor; the gate is equivalent to the control grid of a tube or the base of a regular transistor. For FET symbols, see Fig. 1-3. To emphasize the resemblance of the FET to the tube and to eliminate new terms, some early researchers called the gate, drain, and source by the comparable familiar names of grid, plate, and cathode, but their example failed of adoption.

It is the nature of a PN junction that a thin *depletion layer* is present at the junction. This is a region in which there are no available current carriers. The depletion layer for each FET junction is indicated in Fig. 1-4 as the region inside the dotted lines at the gate junctions. The depletion layer may be deepened by applying a reverse voltage between the gate and source, the depth increasing with voltage. Such an increase in depth narrows the channel through which the current carriers must pass, thus increasing the resistance of that path.

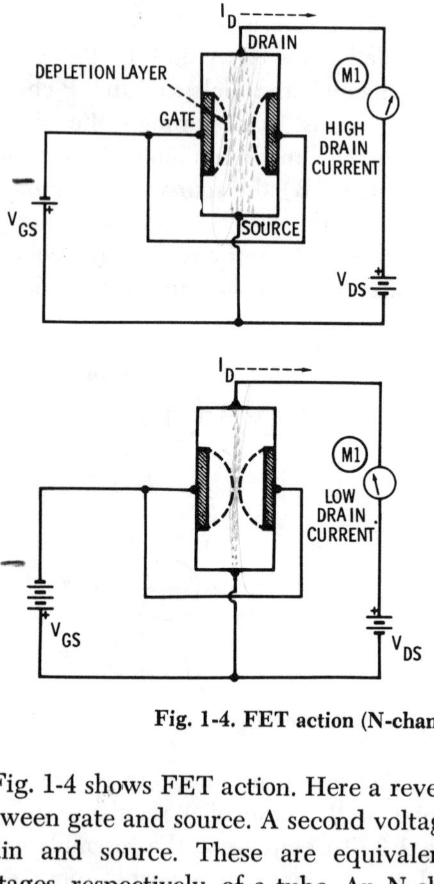

(A) Low gate voltage.

(B) High gate voltage.

Fig. 1-4. FET action (N-channel shown).

Fig. 1-4 shows FET action. Here a reverse voltage, V_{GS}, is applied between gate and source. A second voltage, V_{DS}, is applied between drain and source. These are equivalent to the grid and plate voltages, respectively, of a tube. An N-channel unit is shown; for a P-channel, reverse both V_{GS} and V_{DS}. In Fig. 1-4A, the shallow depletion layers result from the low gate voltage, and the channel between them therefore is wide (permitting a large number of electrons to flow through the bar), so the drain current, I_D, is high. In Fig. 1-4B, the gate voltage is high and it deepens the depletion layers, causing them to penetrate farther into the bar, narrowing the channel and reducing the drain current. When the gate voltage reaches a critical value, termed the *pinch-off voltage,* the depletion layers meet, reducing the current to practically zero. Because the control voltage, V_{GS}, reverse-biases the gate junction, any gate current, I_{GSS}, is exceedingly tiny (on the order of 0.1 nanoampere). This accounts for the high input impedance of the FET and consequently its behavior as a voltage-controlled device. Because the resistance of the silicon bar is modified by the gate-voltage field, the FET is a true field-effect device.

The FET is called *unipolar* from the fact that it uses only one type of current carrier—majority carriers (electrons in the N-channel FET's, and holes in the P-channel). Similarly, the regular transistor is called *bipolar* because it uses both types—majority *and* minority carriers (electrons internal, holes injected, in the NPN; holes internal, electrons injected, in the PNP).

FET PERFORMANCE

The control action in an FET is similar to that in a tube. To illustrate this, Fig. 1-5 shows a family of curves obtained by varying the drain-to-source voltage, V_{DS}, at various levels of gate-to-source voltage, V_{GS}, and noting the drain current, I_D. These curves resemble those of a pentode tube. Note that the avalanche breakdown of the gate junction will be reached if V_{DS} is made high enough. The sudden increase of drain current at the breakdown point is indicated by the dotted extension of each curve. This breakdown voltage decreases as V_{GS} increases, since V_{GS} acts in series with V_{DS} to break down the junction.

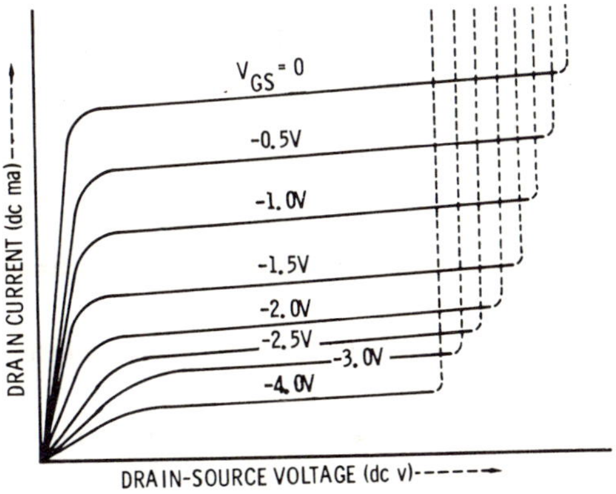

Fig. 1-5. Typical FET performance curves.

Because the FET is a voltage amplifier, its performance, like that of a tube, may be rated in terms of transconductance. In the FET, *forward transconductance* for the common-source circuit (equivalent to the grounded-cathode tube circuit) is the ratio of a change in drain current to the change in gate-to-source voltage which produces it:

$$g_{fs} = \frac{d\,I_D}{d\,V_{GS}} \times 1000$$

where,

g_{fs} is the forward transconductance, in μmhos,
I_D is the drain current, in milliamperes,
V_{GS} is the gate-to-source voltage, in volts.

Depending on make, type, and d-c operating voltages of the FET, the transconductance ranges from a low of 35 to a maximum of 50,000 μmhos.

The FET lends itself readily to use in tube-type circuits. Fig. 1-6, for example, shows a common-source audio-frequency amplifier stage. This is equivalent to the grounded-cathode tube-type amplifier. The following chapters of this book are devoted to the description of such practical circuits.

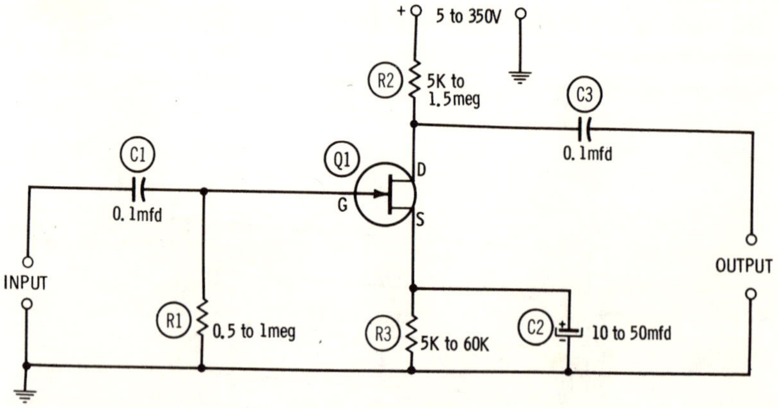

Fig. 1-6. Typical application (FET audio amplifier).

FET RATINGS

In their data sheets, FET manufacturers give minimum and maximum values of various operating parameters. These include the following, dependent on make and type:

1. Common-Source Forward Transconductance (g_{fs}). The ratio of dI_D to dV_{GS}. Similar to tube transconductance (g_m). Given for a specified V_{DS}, V_{GS}, and f = 1 kHz. Range: 35 to 50,000 μmhos.
2. Common-Source Output Conductance (Input Shorted) (g_{oss}). Reciprocal of drain-source output resistance. Given for specified V_{DS}, V_{GS}, and f = 1 kHz. Range: 1 to 600 μmhos.
3. Common-Source Input Capacitance (Output Shorted) (C_{iss}). Capacitance between gate and source. Given for specified V_{DS}, V_{GS}, and f = 1 kHz. Range: 2 to 65 pf.
4. Common-Source Reverse Transfer Capacitance (C_{rss}). Given for specified V_{DS}, V_{GS}, and f = 1 kHz. Range: 1.5 to 6 pf.
5. Drain Current at Zero Gate Voltage (I_{DSS}). The current in the drain-to-source circuit (i.e., through the channel) when V_{GS} = 0. Given for specified V_{DS}. Ranges: V_{DS} = 5 v: 0.1 to 10 ma. V_{DS} = 8 v: 5 to 25 ma. V_{DS} = 10 v: 0.03 to 0.6 ma. V_{DS} = 15 v: 0.2 to 20 ma. V_{DS} = 20 v: 0.4 to 7.5 ma. V_{DS} = 35 v: 80 to 250 ma.
6. Drain Cutoff Current ($I_{D(OFF)}$). Leakage current through the channel when V_{GS} has been adjusted for output cutoff. Given for specified V_{DS} and V_{GS}. Range (for V_{DS} of 15 v): 0.05 na at V_{GS} = 5 v, to 0.07 na at V_{GS} = 10 v.
7. Gate-Drain Voltage (V_{GD}). Also called drain-gate voltage. The maximum voltage which may appear between the gate and drain electrodes. Range: 20 to 50 v at 25°C.
8. Gate Reverse Current (I_{GSS}). Also called gate cutoff current. The reverse current in the gate-to-source circuit. Given for V_{DS} = 0 and a specified V_{GS}. Range: V_{GS} 15 v: 2 na. V_{GS} 20 v: 10 pa to 0.5 na. V_{GS} 30 v: 0.1 to 30 na.
9. Gate-Source Breakdown Voltage (BV_{GSS}). The voltage at which the gate junction will enter avalanche. Given for I_G = 1 μa and V_{DS} = 0. Range: 20 to 50 v at 25°C.
10. Gate-Source Pinch-off Voltage (V_P). The gate-to-source voltage at which the field just closes the conduction channel.

Given for I_D = 1 na, 10 na, or 1 μa, and a specified V_{DS} (e.g., 5 to 15 v). Range: 0.6 to 50 v.

11. Gate-Source Voltage (V_{GS}). Also called source-gate voltage. The maximum voltage which may appear between the gate and source electrodes. Range: 20 to 50 v at 25°C.
12. Noise Figure (NF). Internal noise generated. Given for V_{GS} = 0, and a specified V_{DS} (e.g., 15 v) and frequency (e.g., 1 kHz, 200-Hz bandwidth). Range: 0.5 to 3 db.
13. Total Device Dissipation (P). Maximum power which can be safely dissipated by the FET structure. Range: 200 mw to 0.8 w in free air at 25°C.

FET TYPES

It has already been noted that FET's may be classified as N-channel or P-channel according to whether N-type or P-type semiconductor material is used in the channel. Alternate terminology is NFET and PFET.

In addition to these two basic types, a newer FET employs a metallic gate which is insulated from the semiconductor by a thin oxide film. This device, which provides an extremely high input impedance because of the near-zero I_{GSS} and also gives good r-f performance, is known by four names: IGFET (*insulated-gate field-effect transistor*), MOS (*metal-oxide-semiconductor*), MOS FET (*metal-oxide-semiconductor field-effect transistor*), and MOST (*metal-oxide-semiconductor transistor*). The symbol for this FET is given in Fig. 1-3C.

Another development is the tetrode FET. This unit has two gate terminals ("front gate" and "back gate") and its symbol is given in Fig. 1-3D. The tetrode, which acts like a variable-pinch-off FET, provides a high g_{fs}/C_{gss} ratio and consequent superior performance as a radio-frequency amplifier.

2

Amplifier Circuits

The FET offers several immediate advantages over the bipolar transistor in amplifier circuits. Circuit design techniques, for example, are identical—or very nearly so—with the familiar ones used for tube circuits. And the high input impedance of the FET permits the simple cascading of r-c–coupled stages, without loss of gain between stages. Also, high-impedance devices—crystal microphones and pickups, piezoelectric transducers, and capacitance-type transducers—may be operated directly into FET amplifiers.

Representative amplifier circuits are described in this chapter. The operating data given with them were obtained with the particular FET's used by the author. Therefore, the reader is cautioned that individual performance may vary above or below these figures because of the spread in FET characteristics (the transconductance spread, for example, can be as high as 7.5:1, and the drain current at zero gate voltage can spread over a range of 30:1). The single-stage amplifiers shown here may be cascaded for higher gain.

The input resistance (impedance) of any one of the circuits may be boosted, if desired, by substituting a higher value of gate resistor. In most cases, up to 22 megohms is permissible.

SIMPLE SMALL-SIGNAL A-F PREAMPLIFIER

Fig. 2-1 shows the circuit of a simplified, common-source a-f amplifier employing a 2N3819 FET. The source is grounded directly, and this arrangement is suitable for a-f input levels below 0.6 volt rms.

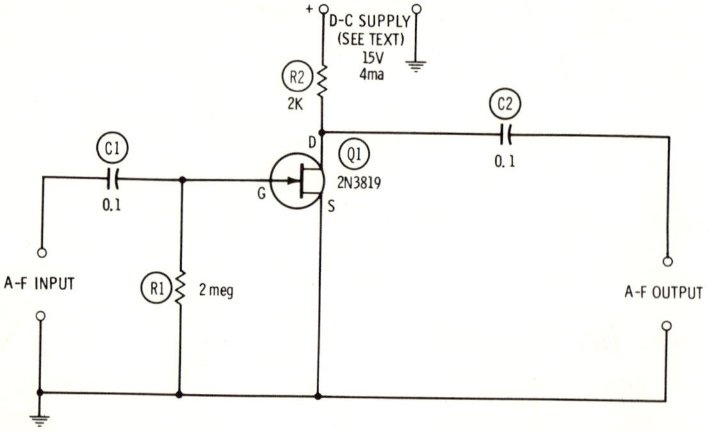

Fig. 2-1. Simple small-signal a-f preamplifier.

At a d-c supply voltage level of 6 volts (2.5 ma), the single stage gives a no-load voltage gain of 2.5. At 9 vdc (3.1 ma), the gain is 3.7. And at 15 vdc (4.0 ma), the gain is 8. The maximum signal input before output-peak clipping is approximately 0.58 volt rms. Frequency response, referred to 1 kHz, is flat within 2 db from 50 Hz to 50 kHz. All gain figures are given on the basis of a 1-megohm output load.

The 2N3819 is encapsulated in plastic and gives no installation problems. All wiring must be kept as short and direct as practicable, for reduction of stray pickup at the low signal levels at which this amplifier can operate.

The 2-megohm gate resistor (R1) and 2000-ohm drain resistor (R2) both are 1/2 watt.

SINGLE-STAGE, LOW-GAIN A-F PREAMPLIFIER

Higher gain than that afforded by the preceding circuit and higher output voltage, as well, may be obtained by self-biasing the

FET. A low-gain circuit of this latter type, based on a 2N4868 FET, is shown in Fig. 2-2. Here, source resistor R3 (bypassed by capacitor C3) provides the required gate bias. Note the similarity of this circuit to that of the familiar grounded-cathode tube amplifier or the common-emitter bipolar-transistor amplifier. All resistors are 1/2 watt, and electrolytic capacitor C3 is rated at 25 dcwv.

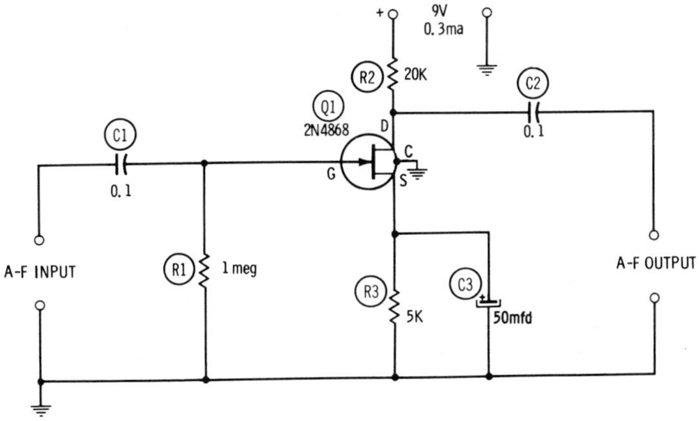

Fig. 2-2. Single-stage, low-gain a-f preamplifier.

With a d-c supply of 9 volts (0.3 ma), the open-circuit voltage gain is 20. The maximum signal input before output-peak clipping is 50 millivolts rms, and the corresponding output signal is 1 volt rms.

All wiring must be kept as short and direct as practicable to minimize stray pickup at the low signal levels at which this amplifier can operate. One pigtail of the 2N4868 is connected to the metal case of this FET and should be grounded, for shielding.

SINGLE-STAGE, MEDIUM-GAIN A-F PREAMPLIFIER

Higher voltage gain than that afforded by the preceding low-gain circuit is obtained with the self-biased, common-source a-f preamplifier, based on a 2N4338 FET, shown in Fig. 2-3. Note here also the similarity of this circuit to that of a grounded-cathode tube amplifier or a common-emitter bipolar-transistor amplifier.

With a d-c supply of 15 volts (8 μa), the voltage gain (open circuit) is 45, and the maximum signal input before output-peak clipping is 78 millivolts rms. This results in a maximum signal output of 3.51 volts rms. Frequency response, referred to 1 kHz, is flat within 0.5 db from 50 Hz to 20 kHz, and within 4 db from 50 Hz to 50 kHz.

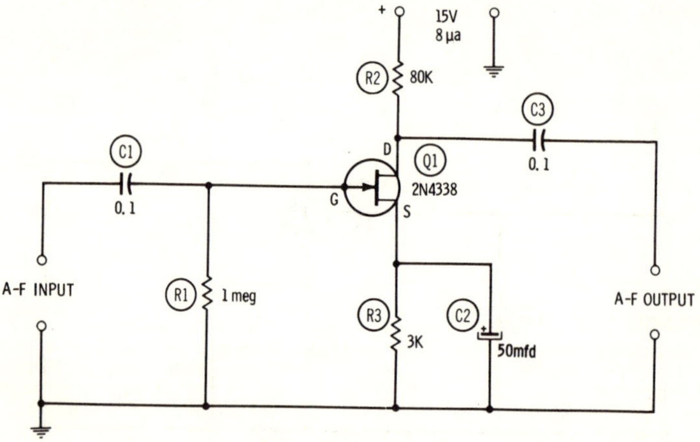

Fig. 2-3. Single-stage, medium-gain a-f preamplifier.

The 1-megohm gate resistor (R1), 80,000-ohm drain resistor (R2), and 3000-ohm source resistor (R3) are 1/2 watt. Electrolytic capacitor C2 is rated at 25 dcwv.

All wiring must be kept as short and direct as practicable, to minimize stray pickup at the low signal levels at which this amplifier can operate. Care must be taken also in the location of the 2N4338, since the gate electrode is internally connected to the metal case of this FET.

A-F SOURCE FOLLOWER

Fig. 2-4 shows the circuit of a source follower employing a 2N4340 FET. Note that with its unbypassed source resistor (R2) across which the output signal voltage is developed, this circuit closely resembles its counterparts—the tube-type cathode follower and bipolar-transistor emitter follower. Like the latter two circuits, the source follower is invaluable for transforming a high-impedance input to a low-impedance output. This circuit finds regular use, therefore, in matching a high-impedance output to a low-impedance

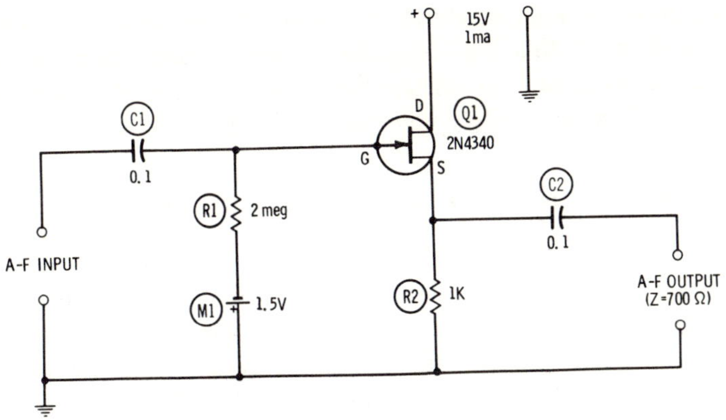

Fig. 2-4. A-f source follower.

transmission line, matching a high-impedance transducer to a low-input-impedance transistor stage, etc. Fixed bias is provided by the 1.5-volt cell, M1. The voltage gain and maximum permissible input-signal voltage are somewhat less when this cell is omitted.

With the d-c supply of 15 volts (1 ma), the voltage gain is 0.5, and the maximum signal input before output-peak clipping is 1.5 volts rms. This results in a maximum signal output of 0.75 volt rms. Frequency response, referred to 1 kHz, is flat within 1 db from 50 Hz to 100 kHz. The gain and frequency response are given here on a no-load basis.

The output impedance of the source follower is somewhat lower than the 1000 ohms of source resistor R2. In this instance, it is 700 ohms. This situation is similar to that which characterizes the cathode follower and emitter follower. In the FET circuit,

$$R_o = \frac{R_s}{1 + R_s\,(g_m + g_o)}$$

where,

R_o is the output impedance (resistance), in ohms,
R_s is the source resistance (R2 in Fig. 2-4), in ohms,
g_m is the forward transconductance of the FET, in mhos,
g_o is the output conductance, in mhos.

All resistors are 1/2 watt, and R1 may be raised as high as 22 megohms if a higher input impedance is desired. Input-circuit wiring must be kept as short and direct as practicable to prevent stray pickup by this high-impedance circuit. Care must be taken also in locating and mounting the 2N4340, since the metal case of this FET is internally connected to the gate electrode.

TWO-STAGE, R-C–COUPLED A-F AMPLIFIER

One of the advantages afforded by the high input impedance of the FET is the latter's adaptability to conventional r-c–coupled multistage amplifiers. As in tube practice, there is no limiting loss of gain between stages as there is in r-c–coupled bipolar-transistor amplifiers. Fig. 2-5 shows the circuit of a 2-stage, r-c–coupled amplifier, employing two 2N4338 FET's (Q1, Q2). Cascaded, self-biased common-source stages are used.

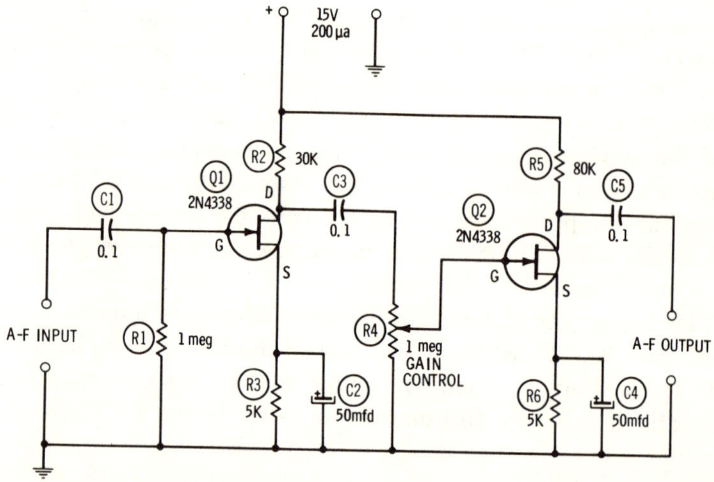

Fig. 2-5. Two-stage, r-c–coupled a-f amplifier.

The total d-c drain is 200 μa at 15 volts. With this d-c supply, the overall voltage gain of the amplifier is 1000 when potentiometer R4 is set for full output. At this level, the maximum signal input before output-peak clipping is 4 millivolts rms, and this corresponds to a maximum signal output of 4 volts rms. Frequency response, referred to 1 kHz, is down 4 db at 50 Hz and at 50 kHz. Gain and frequency-response figures are given in terms of a 1-megohm load across the output.

All resistors are 1/2 watt. Electrolytic capacitors C2 and C4 are rated at 25 dcwv. All wiring must be kept as short and direct as practicable, to minimize stray pickup at low signal levels. Care must be taken also in the location of the 2N4338's, since the gate electrode is internally connected to the metal case of this FET.

TWO-STAGE, TRANSFORMER-COUPLED A-F AMPLIFIER

The high input impedance of the FET makes practicable the use of step-up coupling transformers between amplifier stages, as is common in tube-type amplifiers. The additional voltage gain due to the step-up turns ratio of the transformer thus is easily obtained. (In transformer-coupled bipolar-transistor amplifiers, on the other hand, the low input impedance of the bipolar unit necessitates a step-down ratio.)

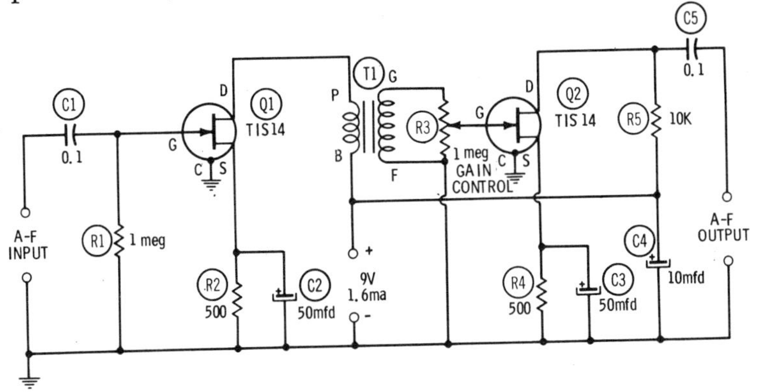

Fig. 2-6. Two-stage, transformer-coupled a-f amplifier.

Fig. 2-6 shows the circuit of a 2-stage amplifier employing transformer interstage coupling. Here, transformer T1 provides a 2:1 step-up, and may be any convenient tube-type interstage transformer. Note the usual primary (P, B) and secondary (G, F) labels. This amplifier is based on two TIS14 FET's (Q1, Q2); however, the same scheme may be used with other field-effect transistors. In fact, some designers will favor transformer coupling to overcome the low voltage amplification provided by low-transconductance FET's.

At 1000 Hz, the voltage gain (with potentiometer R3 set for full output, and the amplifier loaded with 1 megohm) is 3000. This corresponds to a maximum signal input of 1 millivolt rms for 3 volts rms output before output-peak clipping. For high gain, a higher turns ratio than the 2:1 value of the author's transformer may be used. Frequency response of the amplifier is dictated largely by that of the transformer. Total current drain of the circuit is 1.6 ma at 9 vdc.

All leads must be kept as short and direct as practicable, to minimize stray pickup and stray coupling, and the transformer case should be grounded. A fourth pigtail of the TIS14 is connected to the metal shell of this FET and should be grounded, for shielding. All fixed resistors are 1/2 watt, and electrolytic capacitors C2, C3, and C4 are rated at 25 dcwv.

HIGH-IMPEDANCE INPUT FOR BIPOLAR TRANSISTOR

A convenient application of the FET is the provision of high input impedance for a high-gain bipolar-transistor amplifier. In this instance, the FET serves as an input-impedance transformer.

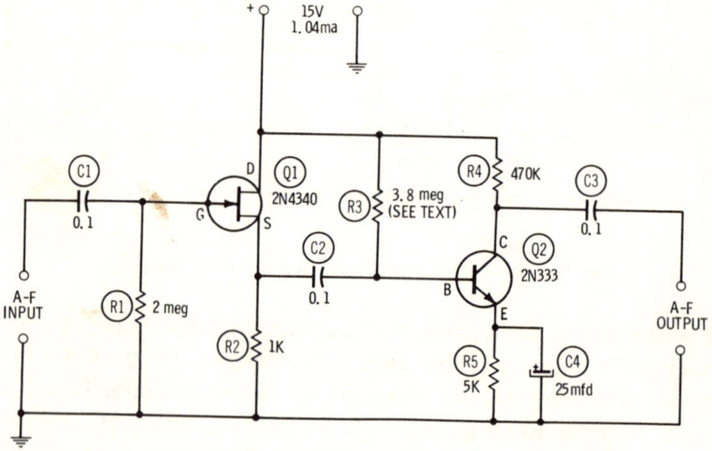

Fig. 2-7. FET high-impedance input for bipolar transistor.

Fig. 2-7 shows such an arrangement. Here, a 2N4340 FET source follower, similar to the one described earlier in this chapter, is connected ahead of a high-gain stage employing a 2N333 silicon transistor (Q2). The 2N333 stage alone provides a voltage gain of

200 when bias resistor R3 is adjusted to an optimum value for the particular transistor used. And since the 2N4340 stage gives a voltage gain of 0.5, the overall gain of the complete circuit is 100 when the output is terminated with 1 megohm.

The maximum signal input before output-peak clipping is 20 millivolts rms, and this corresponds to a signal output of 2 volts rms. The circuit draws approximately 1.04 ma at 15 vdc.

All resistors are 1/2 watt. Electrolytic capacitor C4 is rated at 25 dcwv. Either R1 or R2 may be made a potentiometer for gain control, although varying R2 will introduce less noise.

CONVENTIONAL PHASE INVERTER

Fig. 2-8 shows the circuit of a conventional two-triode a-f phase inverter based on two 2N4338 FET's (Q1, Q2). The two outputs are 180 degrees out of phase with each other. Output 1 is 180 degrees out of phase with the input signal; Output 2 is in phase with the input signal.

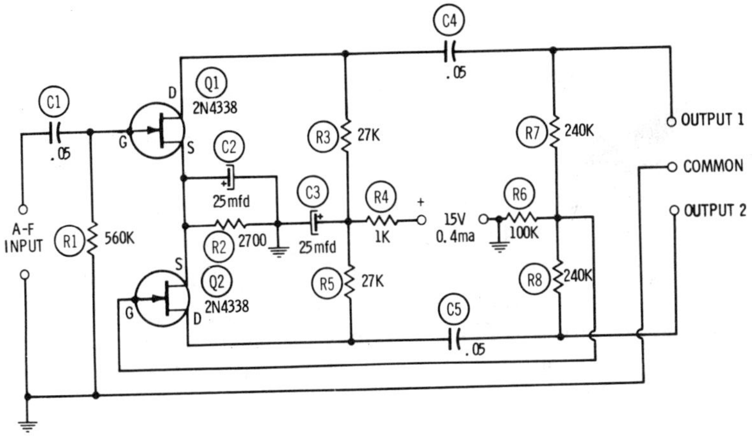

Fig. 2-8. Conventional phase inverter.

The circuit operates in the same manner as the equivalent tube-type circuit. That is, the signal is amplified and inverted by transistor Q1, and the output of Q1 (developed across resistors R7 and R6) is delivered to the OUTPUT 1 terminal. A portion of this output signal is applied to the gate of transistor Q2, and is amplified and inverted by this latter transistor. The output of Q2 (developed across resistors R8 and R6) is delivered to the OUTPUT 2 terminal. Resistors R7 and R8 may serve directly as the gate-to-ground resistors of a succeeding push-pull FET stage, or as the grid resistors if the push-pull stage contains tubes.

Like the equivalent tube circuit, this phase inverter is self-balancing: The signal voltage developed across resistor R6 by the output of transistor Q2 bucks the signal voltage developed across this resistor by the output of Q1 (these two voltages being out of phase with each other), and this action reduces the gate-to-ground signal voltage of transistor Q2 and thus keeps the signal at OUTPUT 1 and OUTPUT 2 substantially equal. The inverter draws 0.4 ma at 15 vdc.

Capacitor C3 and resistor R4 form a decoupling filter which promotes stability; these may be omitted in some layouts if there is no trouble from motorboating. All resistors are 1/2 watt, and electrolytic capacitors C2 and C3 are rated at 25 dcwv. In the 2N4338, the gate electrode is internally connected to the metal case of this FET; hence, care must be taken in the placement of this transistor.

PARAPHASE PHASE INVERTER

A single-FET ("paraphase") phase inverter, equivalent to the single-tube circuit, is shown in Fig. 2-9. This circuit, which employs a U183 FET, exploits the 180-degree phase difference between source output and drain output (equivalent to cathode output and plate output, respectively), and provides the simplest arrangement for converting from single end to push-pull without a transformer.

The signal at the OUTPUT 1 terminal is 180 degrees out of phase with the input signal, that at OUTPUT 2 is in phase with the input. And the equality of drain resistor R2 and source resistor R3, together with the large amount of degeneration supplied by the unbypassed source resistor, keeps these two outputs approximately equal. (They may be equalized exactly through close adjustment of either R2 or R3, one with respect to the other.) The circuit draws 0.25 ma at 15 vdc.

The maximum signal input before output-peak clipping is 1 volt rms, and the voltage gain in each half of the circuit is approximately 0.8. Frequency response, referred to 1 kHz, is flat within 3 db from

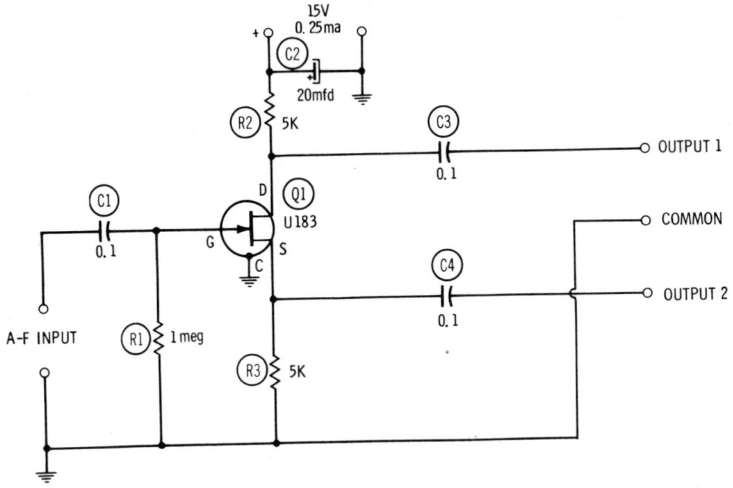

Fig. 2-9. Paraphase phase inverter.

50 Hz to 50 kHz. Gain and frequency-response figures are given for 1-megohm output load.

All resistors are 1/2 watt, and electrolytic capacitor C2 is rated at 25 dcwv. A fourth terminal of the U183 is internally connected to the metal case of this FET and should be grounded, for shielding.

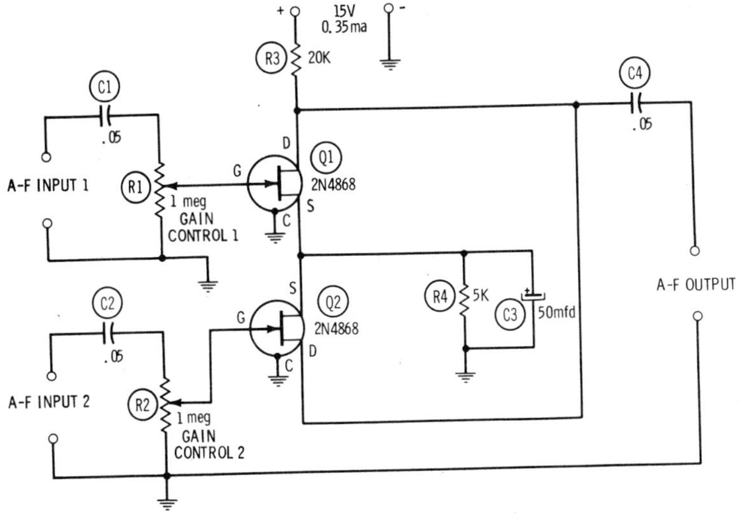

Fig. 2-10. Dual-channel audio mixer.

DUAL-CHANNEL AUDIO MIXER

Fig. 2-10 shows a conventional circuit for mixing two a-f signals or for selecting one or the other. Like the equivalent tube or bipolar-transistor circuit, it provides separate inputs and a common output. Separate gain controls (1-megohm potentiometers R1 and R2) are provided, and the resistances of these controls may be increased to 5 megohms each, if a higher input impedance is desired.

Using two 2N4868 FET's (Q1, Q2), the circuit gives maximum voltage gain of 10 per half. This corresponds to a maximum signal input of 15 millivolts rms for an output of 1.5 volts rms before output-peak clipping. Current drain is 350 μa at 15 vdc.

All resistors are 1/2 watt, and electrolytic capacitor C3 is rated at 25 dcwv. A fourth terminal of the 2N4868 is internally connected to the metal case of this FET and should be grounded, for shielding.

L-C–TUNED BANDPASS A-F AMPLIFIER

An amplifier which can be tuned for peak output at a desired audio frequency is invaluable in bridge balancing, signal selection, telemetering, c-w signal peaking, selective signaling, and electronic control. Fig. 2-11A shows such a circuit which employs two 2N4340 FET's (Q1, Q2) and is tuned by means of inductance and capacitance. Fig. 2-11B shows the frequency response.

The first stage is a common-source amplifier with an unbypassed source resistor (R2). The latter supplies substantial degeneration for stability and low distortion. The second stage is also a common-source amplifier, with separate outputs from its drain (high impedance) and source (low impedance) to accommodate various loads. With potentiometer R3 set for maximum gain, the maximum input signal before output-peak clipping is 7 millivolts rms. The corresponding maximum signal outputs are: A-F OUTPUT 1, 2 volts rms; A-F OUTPUT 2, 0.45 volt rms. The curve in Fig. 2-11B is based upon a 1-megohm load for A-F OUTPUT 1. The circuit draws 0.6 ma at 15 vdc.

The inductance and capacitance values shown in Fig. 2-11A are given for 1-kHz operation: L1 is a U.T.C. Type VI-C15, 5.4-henry adjustable inductor (or equivalent), and C2 is a .005-mfd mica capacitor. A tuning screw allows the inductance to be varied over a

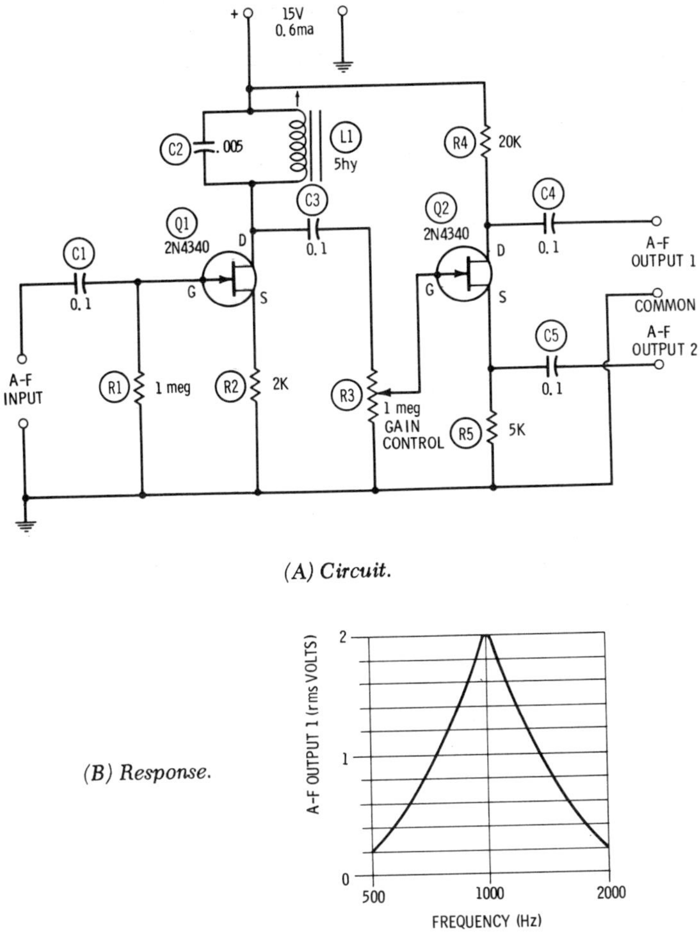

(A) Circuit.

(B) Response.

Fig. 2-11. L-c–tuned bandpass a-f amplifier.

narrow range to set the frequency exactly to 1 kHz. Other operating frequencies may be obtained by changing the value of the coil inductance (L1) or the capacitance (C2). If necessary, both inductance and capacitance can be varied.

All wiring must be kept as short and direct as practicable, to minimize stray coupling and pickup. The metal case of the 2N4340 is internally connected to the gate electrode, so this FET must be mounted clear of contact with other components. All fixed resistors are 1/2 watt.

R-C–TUNED BANDPASS A-F AMPLIFIER

Resistance-capacitance tuning of a bandpass a-f amplifier is a more compact arrangement than the inductance-capacitance tuning described in the preceding section. Fig. 2-12A shows a 1-kHz amplifier which is r-c tuned. Fig. 2-12B shows the frequency response.

The basic amplifier is a 3-stage unit, based on 2N4340 FET's (Q1, Q2, Q3). Unbypassed source resistors (R4, R10, and R13) supply a large amount of current degeneration for stability and

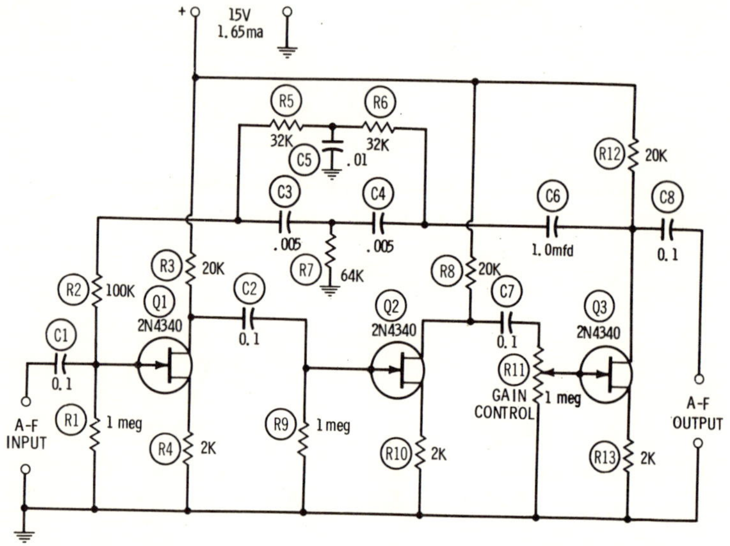

(A) Circuit.

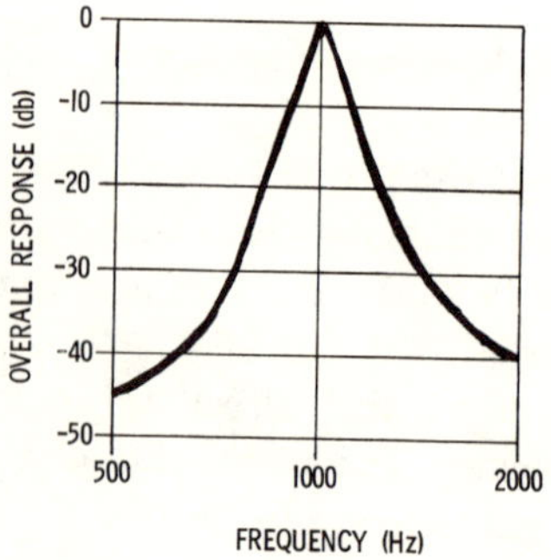

(B) Response.

Fig. 2-12. R-c–tuned bandpass a-f amplifier.

low distortion. The overall voltage gain (with 1-megohm output loading) is 42 db, with potentiometer R11 set for full output, and the odd number of stages rotates the phase correctly for negative feedback in the path through capacitor C6 back to the input.

Tuning is accomplished with a parallel-T network (C3, C4, C5, R5, R6, R7). This is a null network (r-c notch filter) connected in the feedback loop between output transistor Q3 and input transistor Q1. The 1-mfd capacitor, C6, provides d-c blocking for this network, and 100,000-ohm resistor R2 limits the loading of the network. Negative feedback through the loop is sufficient, at the 42-db gain of the amplifier, to cancel the amplifier gain. However, the parellel-T network removes feedback voltage at the network null frequency. The result is transmission by the amplifier very sharply at that frequency, as shown by Fig. 2-12B. The values given for capacitors C3, C4, and C5 and for resistors R5, R6, and R7 are for 1-kHz tuning. Parallel-T networks may be set up for other desired frequencies, provided that the following relationships are maintained:

$$C_3 = C_4 = \frac{C_5}{2}$$

$$R_5 = R_6 = 2R_7$$

The pass frequency of the amplifier then is

$$f = \frac{1}{2\pi CR}$$

where,

f is the pass frequency of the amplifier, in hertz,
C is the capacitance of C3, in farads,
R is the resistance of R5, in ohms.

The amplifier draws 1.65 ma at 15 vdc. All fixed resistors are 1/2 watt. The capacitors and resistors in the parallel-T network must be rated at 1 percent accuracy.

All wiring must be kept as short, rigid, and direct as practicable, for stable operation and minimum pickup of stray signals. The gate electrode of the 2N4340 is internally connected to the metal case of this FET, so the latter should be mounted out of contact with other components.

R-C–TUNED BAND-SUPPRESSION (NOTCH) A-F AMPLIFIER

An amplifier which can be tuned sharply to eliminate an audio frequency is invaluable for suppressing a heterodyne in radiophone reception, eliminating an undesired c-w signal, separating one frequency from a mixture, and removing a single hum or noise component from a signal complex. Fig. 2-13A shows the circuit of a resistance-capacitance–tuned amplifier of this type, based on two

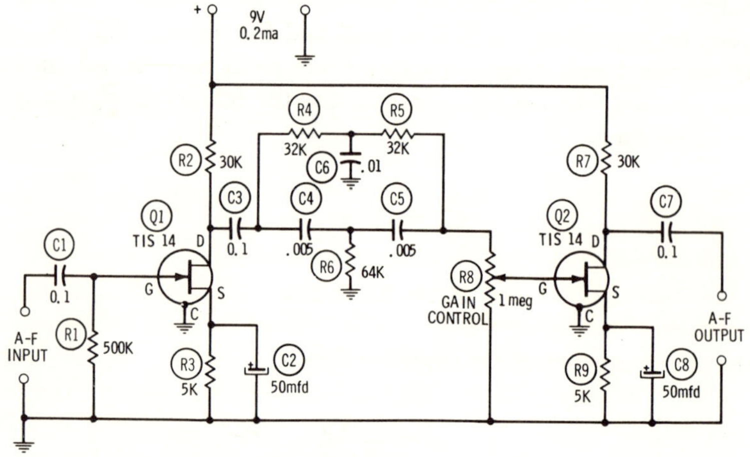

(A) Circuit.

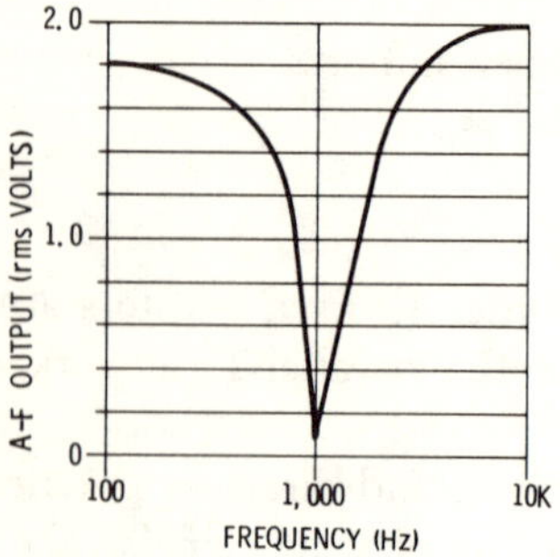

(B) Response.

Fig. 2-13. R-c–tuned band-suppression (notch) amplifier.

TIS14 FET's (Q1, Q2). Fig. 2-13B shows the typical frequency response that can be expected of the circuit. Note how a sharp drop in output of one band of frequencies can be designed into the circuit.

The tuning network is a parallel-T filter (C4, C5, C6, R4, R5, R6) connected between the amplifier stages. The values given for the filter capacitances and resistances in Fig. 2-13A are for 1-kHz elimination. Similar filters may be set up for other frequencies, provided that the following relationships are maintained:

$$C_4 = C_5 = \frac{C_6}{2}$$

$$R_4 = R_5 = 2R_6$$

The null frequency then is

$$f = \frac{1}{2\pi CR}$$

where,

f is the null frequency, in hertz,
C is the capacitance of C4, in farads,
R is the resistance of R4, in ohms.

If the filter capacitors and resistors are closely selected and the capacitors are high Q, the notch point (see Fig. 2-13B) will be very close to zero. At some distance on each side of the notch frequency (say at 0.1f and 10f), the a-f output will be 2 volts rms for a maximum a-f input of 3.75 millivolts rms before output-peak clipping. This assumes that potentiometer R8 is set for maximum gain and that the amplifier is terminated with a 1-megohm resistive load.

All wiring must be kept as short, rigid, and direct as practicable, to promote stability and to minimize stray pickup and stray coupling. A fourth pigtail of the TIS14 is connected to the metal case of this FET and should be grounded, for shielding. All fixed resistors are 1/2 watt, and electrolytic capacitors C2 and C8 are rated at 25 dcwv.

The amplifier draws 0.2 ma at 9 volts dc.

VIDEO AMPLIFIER

Fig. 2-14 shows the circuit of a simple video amplifier using a single 2N3819 FET. This circuit provides a voltage gain of 5: The maximum signal input before output-peak clipping, at 1-megohm load, is 0.6 volt rms for 3 volts rms output. Stages may be cascaded for increased gain.

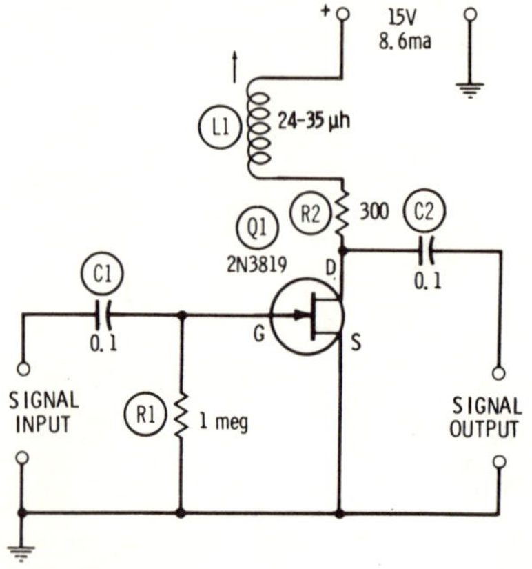

Fig. 2-14. Video amplifier.

The frequency response is within ±2 db from 50 Hz to 4 MHz. Inductor L1 (Miller No. 4508, or equivalent) is adjustable by means of a screw-tuned slug from 24 to 35 μh and is set experimentally for flat output throughout the 50-Hz to 4-MHz range with a constant-amplitude signal applied to the SIGNAL INPUT terminals. Total current drain is 8.6 ma at 15 vdc.

Short, rigid leads are important in this circuit, to minimize strays and to stabilize operation. Resistors R1 and R2 are 1/2 watt.

455-kHz I-F AMPLIFIER

Because of the reverse transfer capacitance of the FET (equivalent to grid-plate capacitance in a tube), the field-effect transistor will oscillate in an amplifier in which l-c tanks in the input (gate) and output (drain) circuits are tuned to the same frequency, unless the circuit is carefully neutralized. Such a circuit is the conventional, transformer-coupled intermediate-frequency amplifier. Neutralization can be a nuisance, since special tapped i-f transformers for single-ended neutralization are not readily available except in the step-down type required by the bipolar transistor.

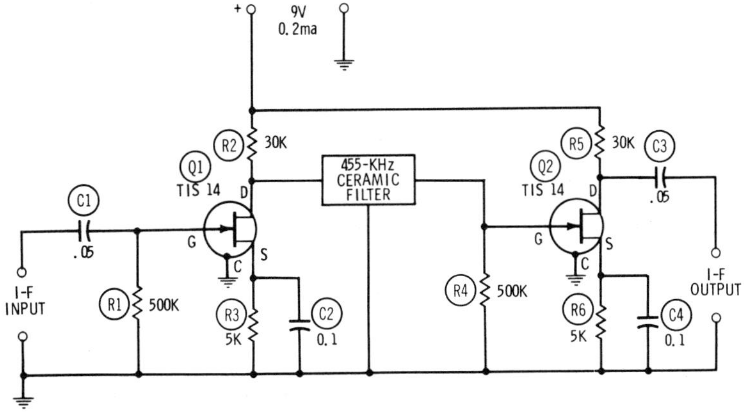

Fig. 2-15. 455-kHz i-f amplifier.

The i-f amplifier shown in Fig. 2-15 avoids this difficulty by using a 455-kHz ceramic filter (U. S. Sonics Type A25H4C, or equivalent) instead of a transformer. This filter, being self-resonant, requires no tuning, and is connected simply between two amplifier stages in place of the usual coupling capacitor. Its voltage insertion loss is approximately 1 db.

The amplifier employs two TIS14 FET's (Q1, Q2) and provides an overall open-circuit gain of 400: The maximum i-f input is 2.5 millivolts rms before peak clipping occurs in the 1-volt rms output. Selectivity of the circuit corresponds very closely to that of the filter alone, i.e., −3 db at 2-kHz bandwidth to −40 db at 160-kHz bandwidth. Total current drain is 0.2 ma at 9 vdc.

All wiring must be kept as short, rigid, and direct as practicable, to minimize i-f losses and to promote stable operation. A fourth pigtail of the TIS14 is connected to the metal case of this FET and should be grounded, for shielding.

AUXILIARY HEADPHONE AMPLIFIER

Often magnetic headphones must be connected to a receiver or other instrument without appreciably loading the device. This requires an auxiliary amplifier with high input impedance.

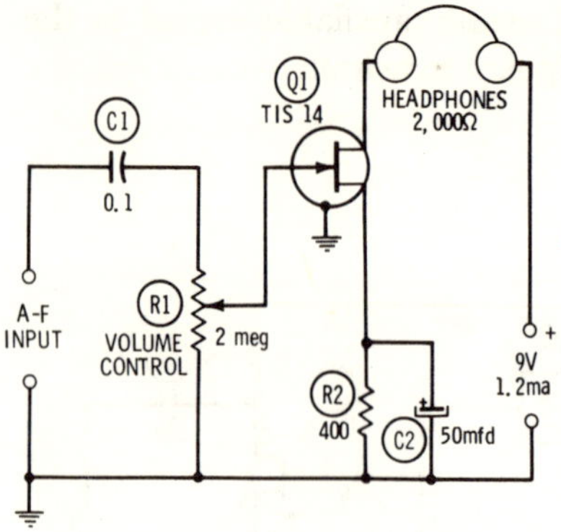

Fig. 2-16. Auxiliary headphone amplifier.

Fig. 2-16 shows the circuit of an amplifier with 2 megohms input resistance, driving a pair of magnetic headphones having 2000-ohms d-c resistance. This amplifier, based on a single TIS14 FET, provides a voltage gain of 40: This corresponds to a maximum a-f signal input of 75 millivolts rms before peak clipping in the 3-volt rms output signal (potentiometer R1 is set for maximum volume).

The circuit draws 1.2 ma at 9 vdc. This power may be supplied by a self-contained battery or may be taken from the power supply of the device which delivers the signal to the headphone amplifier.

Resistor R2 is 1/2 watt, and electrolytic capacitor C2 is rated at 25 dcwv.

SIMPLE AUDIO AGC AMPLIFIER

The gain of the FET amplifier stage is easily controlled by varying the d-c gate bias voltage, the gain varying inversely with the voltage. When the control voltage is derived from some key signal point, such as the output of a multistage amplifier (through rectification and filtering) and applied to one or more stages, automatic gain control results.

Fig. 2-17 shows a simple, one-stage a-f amplifier employing this principle. The d-c control voltage is applied at the CONTROL-VOLTAGE INPUT terminals and acts as additional gate bias for the 2N4868 FET (Q1). Initial operating bias is supplied by the voltage drop across source-bias resistor R4. The maximum current drawn from the control-voltage source is 60 microamperes.

When the d-c control voltage is zero, the voltage gain of the stage is 10, with a maximum undistorted output signal of 1 volt rms. When the control voltage is 6 volts dc, the stage output is reduced to 0.5 millivolt rms (better than 90 db range). A lower d-c control voltage may be used if less db change is needed. For minimum distortion, the maximum a-f input signal reaching the gate of the 2N4868 must be held to 0.1 volt rms. Total current drain is 0.5 ma at 9 vdc.

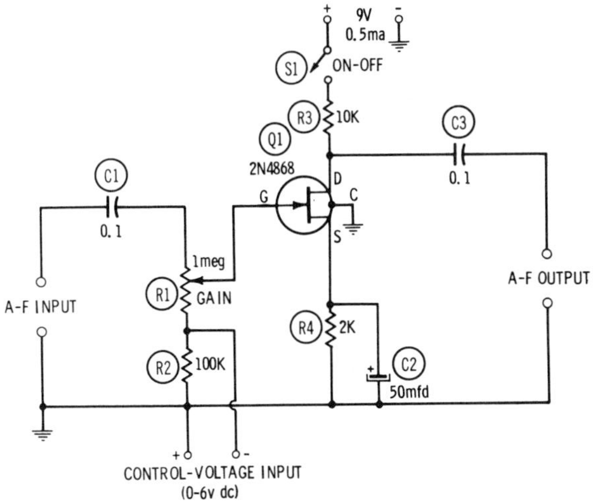

Fig. 2-17. Simple audio agc amplifier.

All fixed resistors are 1/2 watt. The 50-mfd source bypass capacitor, C2, is a 25-dcwv electrolytic, and the 0.1-mfd input and output coupling capacitors, C1 and C3, are 100-volt plastic or paper units. A fourth pigtail of the 2N4868 is connected to the metal case of this FET and should be grounded as shown, for shielding.

The circuit has a great many uses other than conventional automatic gain control in an audio amplifier. Any application is feasible in which an adjustable or fluctuating d-c voltage (at virtually zero power) is available for varying the output of an audio channel. When the control voltage is positive with respect to ground, use a P-channel FET (such as a 2N3578), changing the values of R3 and R4 as required and reversing capacitor C2.

SIMPLE D-C AMPLIFIER

Fig. 2-18 shows the circuit of a simple one-stage d-c voltage amplifier, employing a 2N3578 FET (Q1). This stage has a no-load voltage gain of 5.65. Similar stages may be cascaded for higher gain, provided proper arrangements are made for biasing the gate-source junction of each stage.

When the d-c input signal is zero at the gate of the 2N3578, there is a maximum drain current of almost 1 ma through drain resistor R2. This produces a voltage drop across R2, which reduces the d-c output voltage to approximately 0.35 volt ("false zero"). If exactly zero ouptut is desired, the residual voltage may be eliminated with a common output-bucking circuit. When the d-c input signal is 1.5

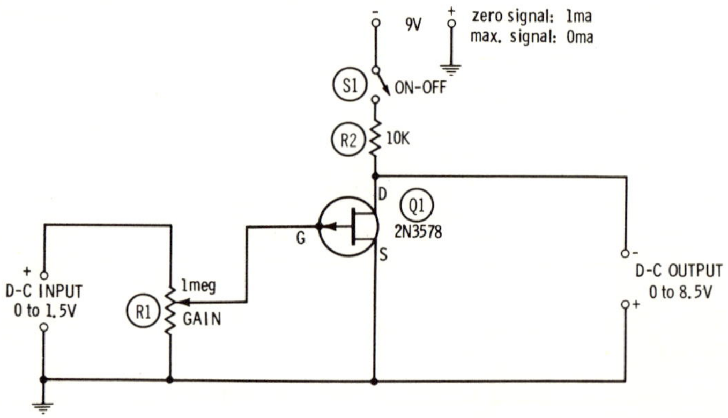

Fig. 2-18. Simple d-c amplifier.

volts at the 2N3578 gate, the drain current is cut off, and the output rises to −8.5 volts. The output values given above are for zero loading. This condition obtains only when the amplifier feeds a high-resistance device, such as an oscilloscope, vtvm, or other amplifier. (At 10,000-ohms load, the output drops to 5 volts dc).

Because a positive input signal gives a negative output signal, the amplifier is an inverter as well, and this feature will be welcome in a number of applications. The unit is also a kind of current amplifier: virtually all of the signal-input current is accounted for in gain-control potentiometer R1, (1.5 μa at 1.5 volts input), whereas the output current flows through an external load as the result of the amplified d-c voltage. This is 500 μa in a 10,000-ohm load—a current gain of 332. (The current and voltage gains referred to here indicate a d-c power gain of 1,000,000 plus).

A simple d-c amplifier of this type is useful in control systems as well as in instruments. For example, it will drive a 10,000-ohm,

5-volt d-c relay, with an input signal of only 1.5 volts at 1.5 microamperes. Various thresholds of operation may be obtained simply by adjustment of gain control R1.

Layout of the amplifier is not critical. However, the 2N3578 must be mounted clear of wiring and other components, as the gate lead is internally connected to the metal case of this FET. Drain resistor R2 is 1/2 watt.

3

Oscillator Circuits

The high input impedance and high gain of the FET promote simplicity and efficiency in many transistorized oscillator circuits. Frequently, the FET may be used directly in tube-type circuits and requires no special circuit components. This latter convenience is important, especially in crystal-controlled, r-c–tuned, and capacitance-feedback oscillators.

Negligible loading of l-c–tuned circuits by the FET may result in higher output and lower distortion than is usually obtained with comparable bipolar transistors. This desirable characteristic also removes the need to tap coils for transistor impedance matching, and tends to preserve undisturbed the Q of the tank circuit. The high gain obtainable with many FET's ensures that sufficient output voltage will be available for efficient feedback.

FET's are proving to be useful in oscillators of all types. They can be used across the frequency-spectrum—from low audio frequencies to high radio frequencies. The oscillator output can be a symmetrical sinusoidal waveform or a nonsinusoidal train of pulses.

TRANSFORMER-FEEDBACK A-F OSCILLATOR

Fig. 3-1 shows the circuit of an audio-frequency oscillator which uses inductance-capacitance tuning and inductive feedback and employs one 2N2608 P-channel FET. This simple arrangement is useful for general-purpose, single-frequency applications, such as bridge excitation, tone signaling, signal injection and tracing, and amplifier testing.

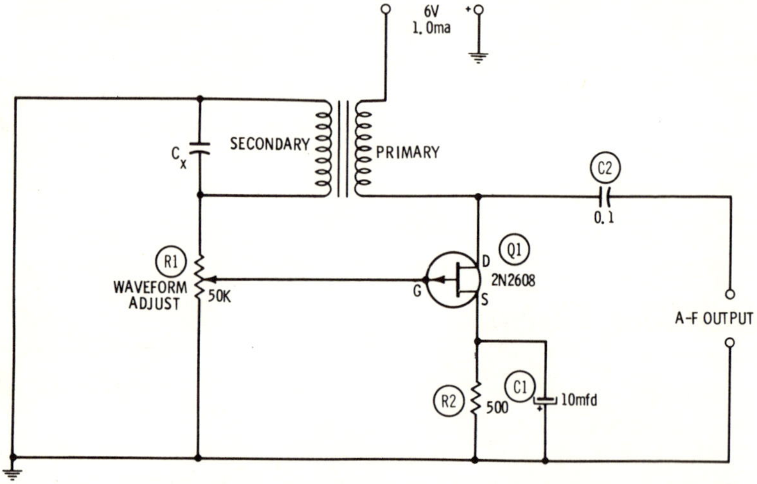

Fig. 3-1. A-f oscillator with transformer feedback.

Feedback from the drain (output) circuit to the gate (input) circuit is provided by an interstage audio transformer, T1, which supplies a 2:1 (or higher) step-up turns ratio between primary and secondary windings. This transformer must be polarized correctly for regenerative feedback, by proper connection of the primary and secondary.

The oscillation frequency is determined by capacitance C_x and the inductance of the transformer secondary:

$$f = \frac{1}{2\pi\sqrt{LC}}$$

where,

f is the oscillation frequency, in hertz,
L is the inductance of the transformer secondary, in henrys,
C is the capacitance of C_x, in farads.

Since transformer manufacturers do not usually specify the inductance of windings, this value must be measured—or various capacitors may be tried as C_x until the desired frequency is obtained. Because of the self-capacitance of the secondary, oscillation will occur at one frequency when no external capacitor is in the circuit. and this is the highest frequency at which the circuit will operate.

The feedback amplitude is adjusted with potentiometer R1. The latter must be set for best sine-wave output shown by an oscilloscope connected to the A-F OUTPUT terminals. Excessive feedback overdrives the FET and clips the output-wave peaks; insufficient feedback causes instability and sluggish starting.

A U.T.C. Type S-8 transformer tested in this circuit gave a frequency of 500 Hz with C_x = .002 mfd, and 1100 Hz with no external capacitor. The open-circuit output voltage was 1 volt rms at 500 Hz, and the current drain 1 ma at 6 vdc. Comparable operation is possible with some of the smaller, transistor-type transformers.

Since the output is capacitance coupled from the drain circuit which includes transformer T1, the external load will tend to detune the oscillator somewhat. This effect can be eliminated, however, by adding a source follower (see Chapter 2) to isolate the oscillator from the load.

The gate electrode of the 2N2608 is internally connected to the metal case of this FET, so care must be taken in the mounting of the FET to avoid contact between the case and other components or wiring. Resistor R2 is 1/2 watt, and electrolytic capacitor C1 is rated at 25 dcwv.

L-C-TUNED A-F OSCILLATOR

In the oscillator circuit shown in Fig. 3-2, the frequency is determined by capacitance C_x and the inductance of iron-cored inductor L1:

$$f = \frac{1}{2\pi\sqrt{LC}}$$

where,

f is the frequency of oscillation, in hertz,
L is the inductance of L1, in henrys,
C is the capacitance of C_x, in farads.

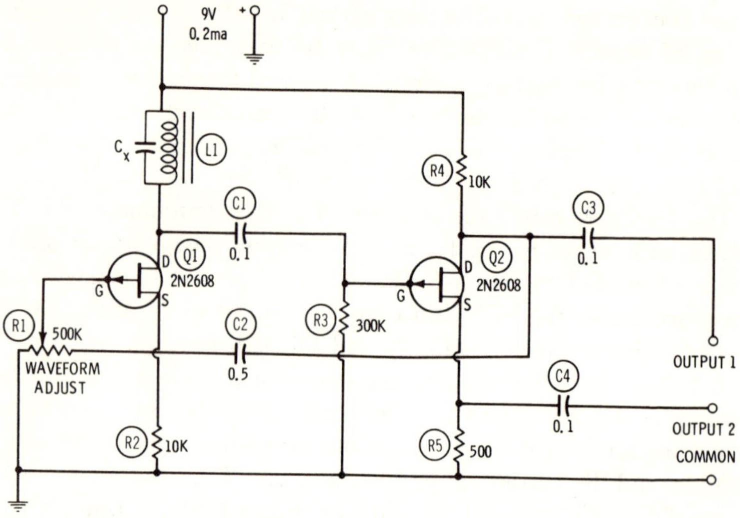

Fig. 3-2. L-c–tuned a-f oscillator.

Here, L may be any convenient coil (from a filter choke to a high-Q adjustable inductor), and C_x may be selected to give a desired frequency. The experimenter may start with a given coil and choose a capacitor, or start with a given capacitor and choose an inductor.

The circuit is essentially a 2-stage r-c–coupled amplifier, employing 2N2608 FET's, with the tuned circuit (LC_x) forming the plate tank of the first stage, and with overall feedback for oscillation provided by capacitor C2. An even number of stages is necessary for the required positive feedback.

Potentiometer R1 allows the feedback amplitude to be adjusted for best sine-wave output, as observed with an oscilloscope connected to the output terminals (either OUTPUT 1 or OUTPUT 2). Excessive feedback overdrives Q1 and distorts the output wave; insufficient feedback causes instability and sluggish starting.

With L1 = 5 hy and C_x = .005 mfd, the frequency is 1000 Hz. The open-circuit signal voltage at OUTPUT 1 (high-impedance output) is 1 volt rms, and at OUTPUT 2 (low-impedance output) is 0.6 volt rms. Current drain is 0.2 ma at 9 vdc.

All wiring must be kept short, direct, and rigid for maximum stability and minimum stray pickup, and the entire device should be box shielded. The gate electrode of the 2N2608 is internally connected to the metal case of this FET, so care should be taken to avoid contact between the FET's and other components. Fixed resistors R2, R3, R4, and R5 are 1/2 watt.

PHASE-SHIFT A-F OSCILLATOR

The phase-shift oscillator is well known for its excellent sine-wave output and the simplicity and compactness of the resistance-capacitance network which determines its frequency. Fig. 3-3 shows an FET version of this familiar circuit, employing a single 2N4338.

The frequency-determining r-c network consists of identical ca-

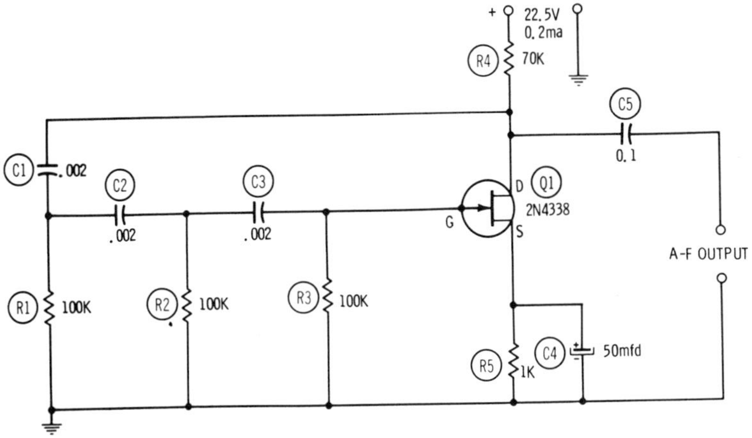

Fig. 3-3. Phase-shift a-f oscillator.

pacitors C1, C2, and C3, and identical resistors R1, R2, and R3. Each leg of this network (i.e., C1-R1, C2-R2, and C3-R3) produces a 60-degree phase shift. The total 180-degree phase shift thus introduced in the feedback path from drain to gate of the 2N4338 is correct for the positive feedback needed for oscillation. Oscillation occurs at that frequency at which the phase angle of each leg of the network is 60 degrees:

$$f = \frac{1}{10.88\ RC}$$

where,

f is the frequency of oscillation, in hertz,
R is the resistance of R1, R2, or R3, in ohms,
C is the capacitance of C1, C2, or C3, in farads.

From this relationship, C = 1/(10.88fR) and R = 1/(10.88fC). All capacitors and resistors must be rated at 1 percent or better accuracy.

The network values given in Fig. 3-3 (C1 = C2 = C3 = .002 mfd, and R1 = R2 = R3 = 100,000 ohms) give a signal frequency of approximately 460 Hz. The open-circuit output voltage of the circuit at this frequency is 8 volts rms. Current drain is 0.2 ma at 22.5 vdc.

In this single-stage circuit, the FET must be a strong amplifier (i.e., its transconductance must be high), or no oscillation will be obtained. Specifically, the voltage amplification of the circuit must be high enough to override the insertion loss of the r-c network; otherwise, the feedback voltage reaching the gate will be insufficient to set up oscillation.

As in other simple oscillators, the output here is capacitance coupled from the drain circuit which includes the input end of the phase-shift network. Because of this, the external load may tend to detune the oscillator somewhat. This defect may be corrected, however, by adding a source follower (see Chapter 2) to isolate the oscillator from the load.

All wiring must be kept short, rigid, and direct for maximum stability and minimum stray pickup. However, complete shielding is unnecessary unless the oscillator is operated in an environment of strong magnetic fields. But care must be taken to mount the 2N4338 clear of other components and wiring, as the gate electrode is internally connected to the metal case of this FET and is vulnerable to capacitive pickup, as well as to short circuits and grounds. All resistors are 1/2 watt, and electrolytic capacitor C4 is rated at 25 dcwv.

SINGLE-FREQUENCY WIEN-BRIDGE A-F OSCILLATOR

Another widely used r-c–tuned audio oscillator is the Wien-bridge type. This circuit is somewhat simpler than the phase-shift oscillator described in the preceding section, since the Wien bridge uses only two resistors and two capacitors to set the frequency, whereas the phase-shift oscillator requires three capacitors and three resistors. However, the Wien-bridge circuit requires two FET's.

Fig. 3-4 shows a Wien-bridge oscillator circuit arranged for single-frequency audio operation. This circuit, employing two

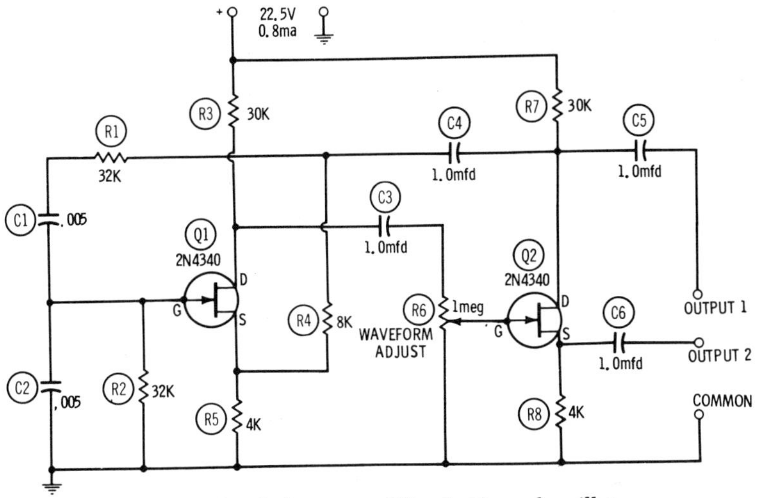

Fig. 3-4. Single-frequency Wien-bridge a-f oscillator.

2N4340 FET's (Q1, Q2), is essentially a 2-stage r-c–coupled amplifier with a Wien bridge (C1, C2, R1, R2, operated against R4, R5) connected in the feedback path from the output drain (Q2) to the input gate (Q1). In this frequency-determining r-c network, C1 = C2, and R1 = R2. At one frequency, determined by these R and C values, the feedback voltage at the Q1 gate is in the proper phase for oscillation, and at all other frequencies degeneration produced by the feedback-voltage drop across R4 + R5 cancels oscillation. The output of the Wien-bridge oscillator therefore has an excellent sinusoidal waveform. When C1 = C2, and R1 = R2,

$$f = \frac{1}{2\pi RC}$$

where,

f is the oscillation frequency, in hertz,
R is the resistance of R1, in ohms,
C is the capacitance of C1, in farads.

From this relationship, C1 = $1/(2\pi fR)$, and R1 = $1/(2\pi fC)$.

The C1, C2, R1, and R2 values given in Fig. 3-4 are correct for 1000-Hz operation. Excellent sine-wave output is obtained at this frequency when potentiometer R6 is set for the best waveform observed with an oscilloscope connected to the OUTPUT 1 and COMMON terminals. The no-load output is 6 volts rms at the OUTPUT 1 (high-impedance) terminal, and 0.4 volt rms at the OUTPUT 2 (low-impedance) terminal. Current drain is 0.8 ma at 22.5 vdc.

For maximum stability and minimum stray pickup, all wiring must be kept short, rigid, and direct. And since the gate electrode of the 2N4340 is internally connected to the metal case of this FET, both transistors must be mounted clear of other components. All fixed resistors are 1/2 watt. Since the high output is capacitance coupled from the Q2 drain circuit which is also coupled to the Wien-bridge network through capacitor C4, the output load may tend to detune the oscillator somewhat. However, this defect may be corrected, as recommended for the phase-shift oscillator, by adding a source follower (see Chapter 2) to isolate the oscillator from the load.

DRAIN-COUPLED MULTIVIBRATOR

Fig. 3-5 shows the circuit of a free-running multivibrator, based on two 2N4340 FET's (Q1, Q2). This arrangement is equivalent to the plate-coupled, tube-type multivibrator.

The multivibrator is essentially an r-c–coupled amplifier with the output of each of two stages driving the input of the other. Thus, in Fig. 3-5 the drain of Q1 is coupled through capacitor C1 to the

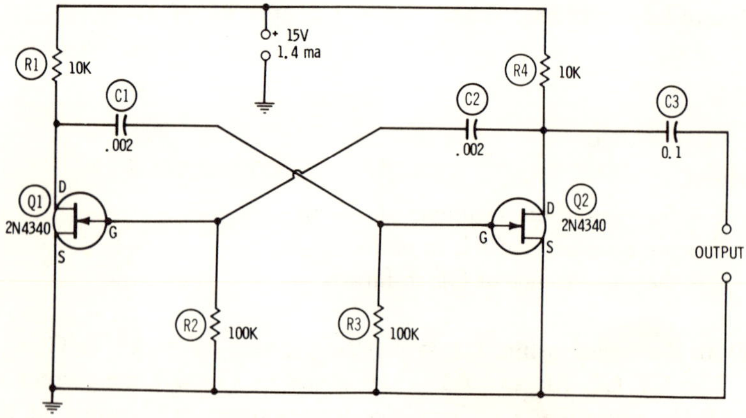

Fig. 3-5. Drain-coupled multivibrator.

gate of Q2, and the drain of Q2 is coupled through capacitor C2 back to the gate of Q1.

As in the tube circuit, the generated frequency is governed by the time constants of the cross-coupled r-c legs:

$$f = \frac{1}{C_1R_3 + C_2R_2}$$

where capacitances are in farads and resistances in ohms. If the components are closely matched (i.e., C1 = C2, and R2 = R3), the formula becomes simply

$$f = \frac{1}{2\,C_2R_2}$$

The coupling resistance and capacitance values in Fig. 3-5 give 2500-Hz operation. Some adjustment of these values may be required with individual FET's.

The output signal is a rectangular wave having a no-load amplitude of 12 volts peak to peak when a 15-volt d-c supply is used. Current drain is 1.4 ma. A synchronizing signal may be injected at either gate or either drain.

Resistors R2 and R3 must match within 1 percent or better, and so must capacitors C1 and C2. All resistors are 1/2 watt. Wiring should be kept short and rigid, but lead dress is not critical except at frequencies of 5 kHz and higher. The 2N4340's should be installed out of contact with each other or with other components, since the gate electrode is internally connected to the metal case in this FET.

SOURCE-COUPLED MULTIVIBRATOR

A somewhat simpler free-running multivibrator circuit than the one described in the preceding section is shown in Fig. 3-6. This source-coupled multivibrator is equivalent to the cathode-coupled, tube-type circuit. In this arrangement, using two 2N4340 FET's (Q1, Q2), only one interstage coupling capacitor (C1 in the forward path from Q1 to Q2) is needed. The coupling back from the second FET (Q2) to the first (Q1) is obtained through the common source resistor, R3.

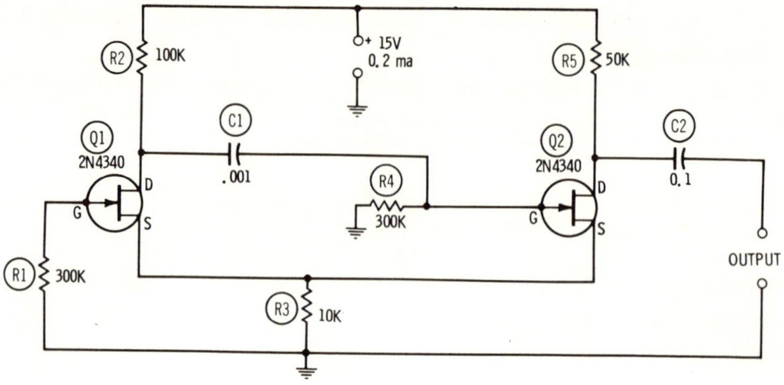

Fig. 3-6. Source-coupled multivibrator.

The generated frequency depends principally upon capacitance C1 and resistance R4. The C1 and R4 values shown in Fig. 3-6 are for 550-Hz operation. Increase C1 for lower frequencies, and decrease it for higher frequencies. The output is a rectangular wave with an amplitude of 10.5 volts peak to peak when the d-c supply is 0.2 ma at 15 volts dc. A synchronizing voltage may be injected at the gate of Q1.

All resistors are 1/2 watt. As in the previous circuit, wiring should be kept short and rigid, but lead dress is not critical except at frequencies of 5 kHz and higher. The 2N4340's must be installed free from contact with each other or with other components, since the gate electrode is internally connected to the metal case in this FET.

SELF-EXCITED R-F OSCILLATOR

The 2N3823 FET has an excellent high-frequency figure of merit (ratio of transconductance to input capacitance), hence is useful for r-f applications through the vhf region. Fig. 3-7 shows the circuit of a self-excited radio-frequency oscillator employing this FET.

This is a series-fed Hartley circuit, in which the tank circuit L1-C1 determines the generated frequency. The values of C1 and L1 may be determined for a desired frequency, either in terms of an available capacitance or available inductance:

$$C_1 = \frac{1}{39.5\, f^2 L_1}$$

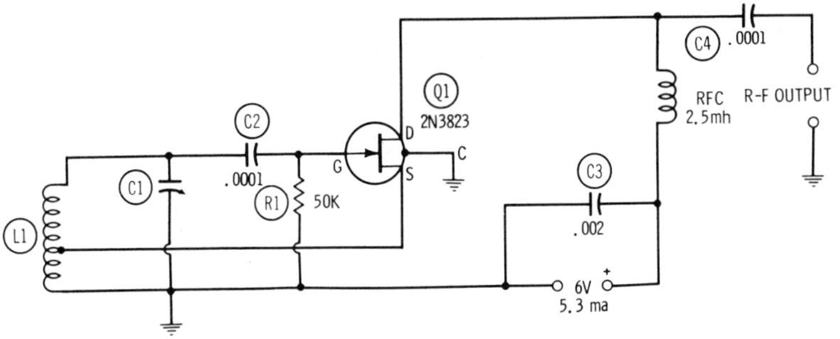

Fig. 3-7. Self-excited r-f oscillator.

$$L_1 = \frac{1}{39.5\ f^2C_1}$$

where,

C_1 is capacitance, in microfarads,
L_1 is inductance, in microhenries,
f is frequency, in megahertz.

The inductance and capacitance values may also be found with the aid of a reactance slide rule or l-c-f chart. For variable-frequency operation, either C1 or L1 may be adjustable. The coil is center tapped, but in some instances the tap may have to be displaced somewhat from center.

Radio-frequency output may be capacitance coupled from the drain, as shown in Fig. 3-7. In this case, the no-load r-f output amplitude is 2.5 volts rms when the d-c input is 5.3 ma at 6 vdc. Link coupling out of L1 also may be employed—the conventional manner. The r-f potential across L1 is 1.3 volts peak.

All wiring must be kept short, direct, and rigid, especially at frequencies above 1 MHz. The case of the 2N3823 is connected to a fourth pigtail which should be grounded, as shown. All resistors are 1/2 watt.

CONVENTIONAL CRYSTAL OSCILLATOR

Fig. 3-8 shows the circuit of a crystal-controlled r-f oscillator employing a 2N3823 FET. This circuit is equivalent to the conventional plate-tuned, tube-type circuit and is operated in the same way as the latter.

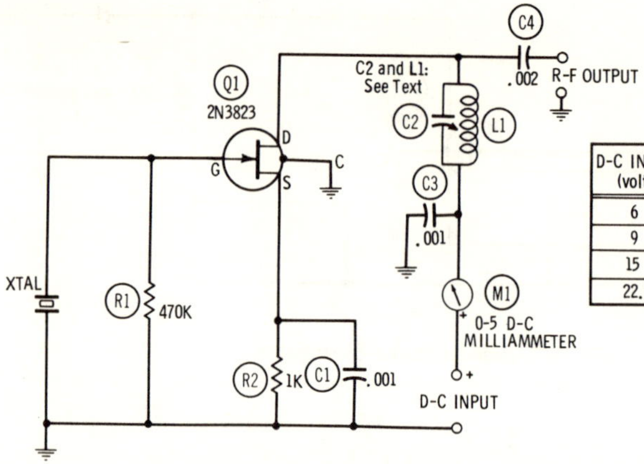

D-C INPUT (volts)	M1 CURRENT (ma)	R-F OUTPUT (v rms)
6	1.40 ma	0.40
9	1.41	0.52
15	2.00	0.80
22.5	2.50	1.40

Fig. 3-8. Conventional crystal oscillator.

In this arrangement, capacitance C2 and inductance L1 are chosen to resonate at the crystal frequency. These tank-circuit values may be determined for a desired frequency either in terms of an available capacitance or available inductance:

$$C_2 = \frac{1}{39.5\, f^2 L_1}$$

$$L_1 = \frac{1}{39.5\, f^2 C_2}$$

where,

C_2 is capacitance, in microfarads,
L_1 is inductance, in microhenries,
f is frequency, in megahertz.

The inductance and capacitance values may be determined also with the aid of a reactance slide rule or l-c-f chart. Either C2 or L1, or both, may be adjustable. The circuit is set to the crystal frequency in the conventional manner—by tuning either C2 or L1 for drain-current dip indicated by d-c milliammeter M1. Fig. 3-8 shows the no-load r-f output voltages obtained with various values of d-c supply voltage and current.

A simple capacitance-coupled r-f output is shown in Fig. 3-8. However, energy may also be inductively coupled out of the oscillator, with a link coil coupled to the lower end of coil L1.

All wiring must be kept short, direct, and rigid. The case of the 2N3823 is connected to a fourth pigtail which should be grounded, as shown. Both resistors are 1/2 watt. For best high-frequency performance, C1, C3, and C4 should be mica capacitors; and if C2 is fixed, it should be silvered mica.

PIERCE CRYSTAL OSCILLATOR

The Pierce crystal oscillator is easy to use, since it requires no tuning. This circuit has a long record of convenience in calibrators, marker generators, crystal-testing oscillators, and radio transmitters.

Fig. 3-9 shows a Pierce circuit and the voltages obtained with various values of d-c supply. This is a vigorous oscillator.

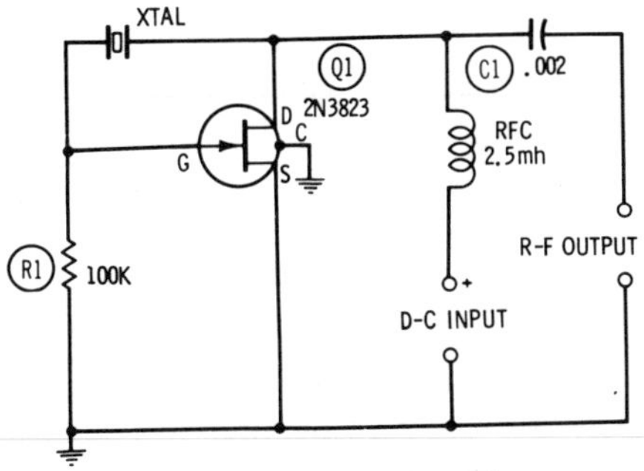

D-C INPUT (volts)	INPUT CURRENT (ma)	R-F OUTPUT (v, peak)
6	1.8	5
9	2.0	6.6
15	2.6	11.0

Fig. 3-9. Pierce crystal oscillator.

The operator must be familiar with the type of crystal he uses in this circuit, as the Pierce oscillator favors the fundamental frequency. Harmonic-type crystals operate here at their *fundamental* frequency only.

All wiring must be kept short and rigid. The case of the 2N3823 is connected to a fourth pigtail which should be grounded, as shown. Resistor R1 is 1/2 watt, and capacitor C1 should be a mica unit.

100-kHz CRYSTAL OSCILLATOR

A 100-kHz crystal oscillator is invaluable for use as a frequency standard and crystal calibrator. Fig. 3-10 shows the circuit of a simple oscillator of this type, employing a 2N3823 FET, which requires no tuning.

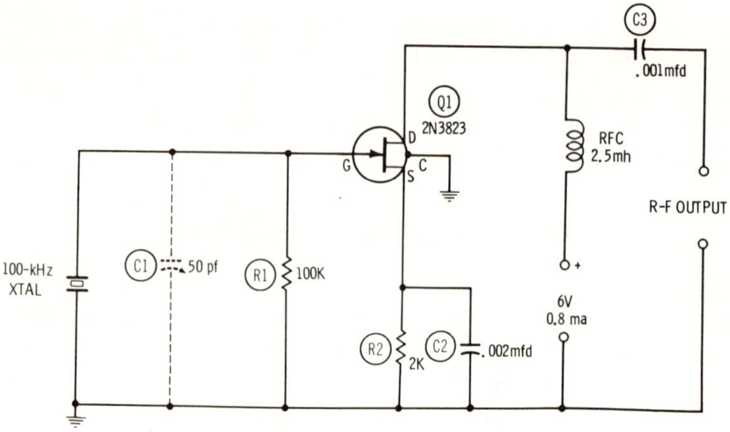

Fig. 3-10. 100-kHz crystal oscillator.

The circuit is a fast starter and vigorous oscillator. At 6 volts, 0.8 ma d-c input, the no-load r-f output is 5.4 volts rms, and this output is sufficiently nonsinusoidal that the 100-kHz harmonics may be detected far into the radio-frequency spectrum. If the oscillator frequency is to be set against standard-frequency signals from WWV or some other standard, the optional air trimmer capacitor C1 may be added. A 50-pf unit should allow the crystal frequency to be shifted sufficiently for the zero beating of a suitable harmonic with the high-frequency standard signal.

All wiring must be kept short, direct, and rigid. The metal case of the 2N3823 is connected to a fourth pigtail which should be grounded, as shown. Resistors R1 and R2 are 1/2 watt, and C2 and C3 are mica capacitors. (The capacitance of C3 may need to be reduced in some applications to minimize loading of the oscillator. In all instances, coupling should be as loose as possible.)

COLPITTS A-F OSCILLATOR

The Colpitts oscillator has the advantage that its inductor needs no tap, and no transformer is required for feedback. In spite of this simplicity, the oscillator is capable of a good output waveform and will operate at radio and audio frequencies.

Fig. 3-11 shows the circuit of a Colpitts a-f oscillator. While this is essentially a single-frequency unit (900 Hz), the frequency may be varied over a narrow range (650-900 Hz) if a slug-tuned, 5-

henry inductor (United Transformer VIC-15, or equivalent) is used as shown in Fig. 3-11.

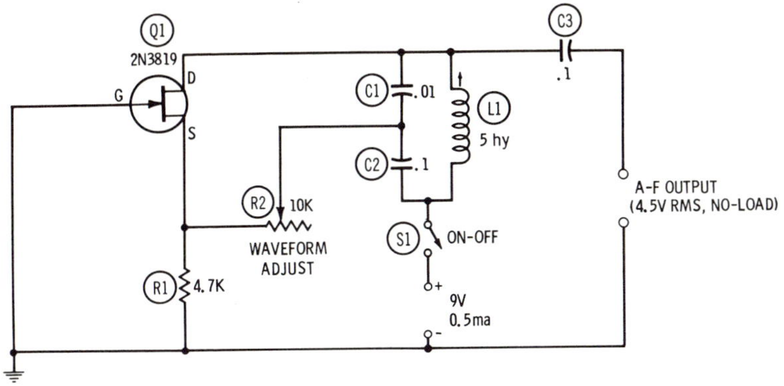

Fig. 3-11. Colpitts a-f oscillator.

This oscillator is built around a single plastic-encapsulated 2N3819 FET (Q1). Direct-current drain is 0.5 ma at 9 volts, and the corresponding no-load output signal is 4.5 volts rms. With the aid of an oscilloscope or distortion meter connected to the A-F OUTPUT terminals, rheostat R2 may be set for best sine waveform.

The frequency may be changed by substituting other values for the C1 and C2 capacitances shown here; however, the 10:1 ratio must be preserved between C1 and C2. A small shift of frequency may be obtained, when a fixed inductor is used for L1, by varying only one capacitance. If, for example, L1 = 5 hy and C2 is changed to .07 mfd, the frequency becomes 1 kHz. However, rheostat R2 must then be readjusted for the best output waveform.

Source resistor R1 is 1/2 watt. Frequency-determining capacitors C1 and C2 must be high-grade units, for stability and good waveform. (While mica is desirable, such capacitors are expensive in the large capacitances shown here; good plastic units are almost as satisfactory.) Output coupling capacitor C3 is a 100-volt plastic unit.

CODE-PRACTICE OSCILLATOR

An interesting application of the Colpitts circuit of the previous section is the code-practice oscillator shown in Fig. 3-12. In this unit, the magnetic headphones themselves supply the inductance of the frequency-determining circuit, and no other coil or transformer is needed.

This oscillator, which employs a single plastic-encapsulated 2N3819 FET (Q1), gives a hefty signal (a maximum output of 4.2 volts rms is developed across a pair of Trimm 2000-ohm magnetic headphones). Current drain is 0.8 ma at 9 vdc.

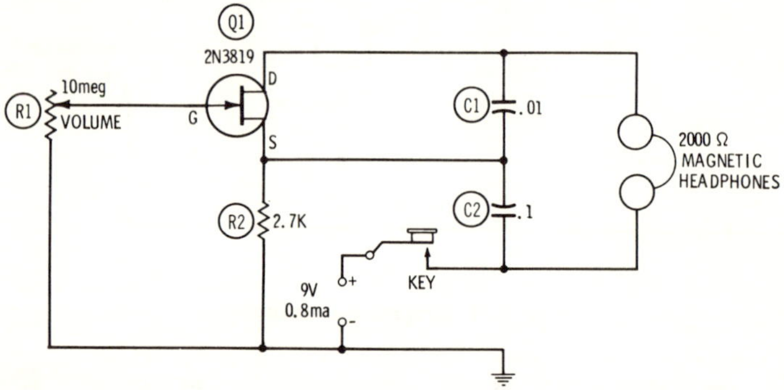

Fig. 3-12. Code-practice oscillator.

With the Trimm phones and with C1 = .01 mfd and C2 = 0.1 mfd, the signal frequency is approximately 1000 Hz. This can be changed by altering the C1 and C2 values while preserving their 10:1 capacitance ratio; however, limited frequency variation may also be obtained, without killing oscillation, by varying only C2. The lower the capacitances, the higher the frequency is and vice versa.

The oscillator keys cleanly and has a good smooth tone. The headphone volume may be varied, with only slight effect on the frequency and waveform, by means of 10-megohm rheostat R1. Since there is no current unless the key is depressed, there is no need for an ON-OFF switch.

Resistor R2 is 1/2 watt. The quality of capacitors C1 and C2 is not critical in this application, so both capacitors may be of any type that is convenient to the user. However, if frequency stability and excellent waveform are demanded, C1 and C2 should be good grade 100-volt plastic units.

4

Receiver and Receiver-Accessory Circuits

Use of the field-effect transistor presents few problems in conventional receiver circuits. In fact, some high-voltage FET's, such as Crystalonics Type C632, will replace tubes directly at the original plate voltages if cathode resistors are changed (where this is required) and i-f and r-f stages are neutralized when these stages oscillate with triodes. Lower-voltage FET's may be employed readily in receiver circuits set up especially for their use.

The receiver circuits given in this chapter are of especial interest to the hobbyist and experimenter. In addition to actual receivers, some accessory devices for use with receivers are also described.

SUPERHET BROADCAST RECEIVER

Fig. 4-1 shows the circuit of a superheterodyne broadcast-band receiver (540-1650 kHz). Three 2N3823 FET's are employed in the combined oscillator-converter (Q1) and first and second r-f amplifiers (Q2, Q3); a TIS14 FET (Q4) is used in the a-f driver stage. The second detector is a 1N295 (X1), and the push-pull, class-B a-f output stage employs 2N169A bipolar transistors (Q5, Q6).

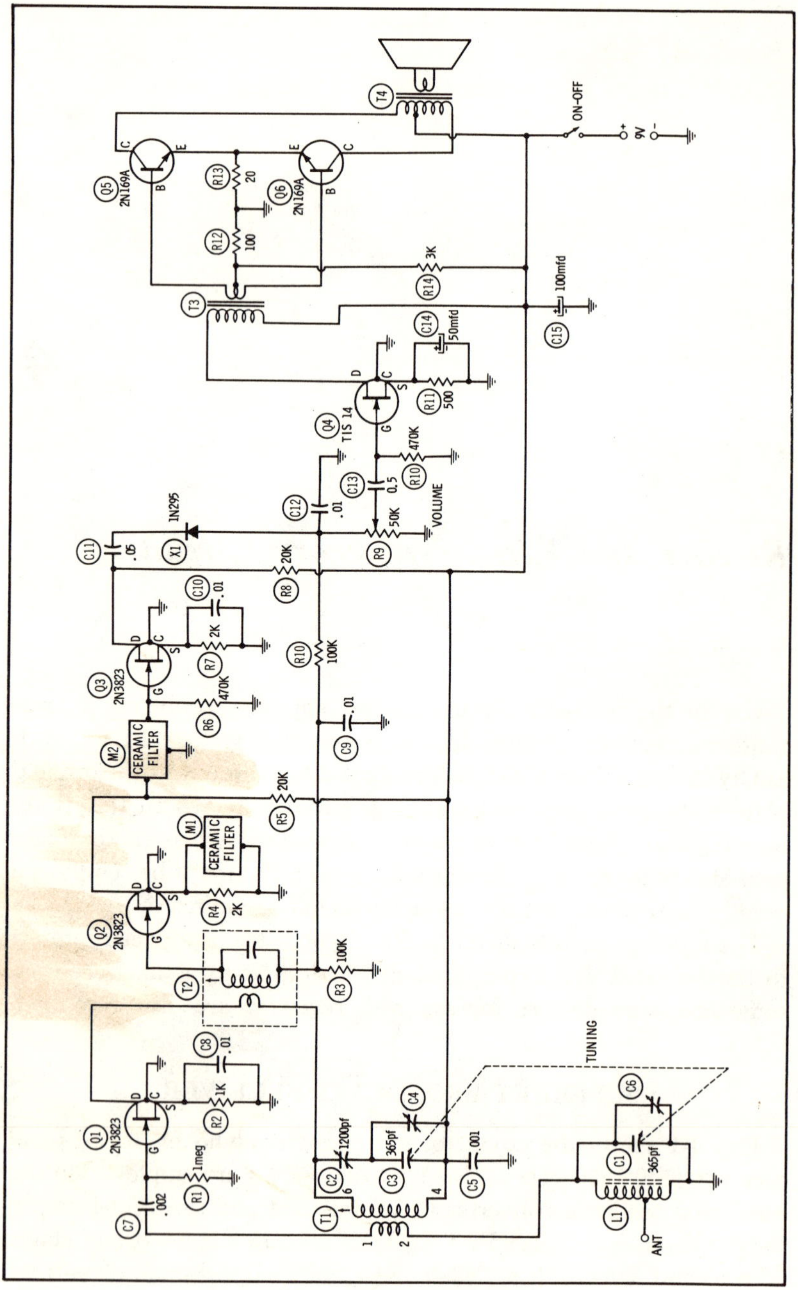

Fig. 4-1. Superhet broadcast receiver.

The self-contained ferrite antenna, L1, is a Miller 2004 (or equivalent). Because the FET produces negligible loading, Q1 is connected across this entire antenna coil, and an outside antenna (when used) is connected to the low-impedance tap (ANT), giving a signal step-up through the autotransformer thus created. The oscillator transformer, T1, is a Miller 2020 transistor-type unit (or equivalent). Tuning throughout the standard-broadcast band is accomplished with dual 365-pf variable capacitor C1-C3. Trimmer capacitors C4 and C6 are part of the tuning capacitor, but C2 is a separate padder with 1200-pf maximum capacitance. During initial alignment of the receiver, the antenna and oscillator are trimmed by means of C6 and C4, respectively, and the oscillator adjusted additionally with the aid of the tuning slug of T1.

The converter is coupled to the i-f amplifier through a conventional transistor-type i-f transformer, T2, which is reverse connected, as shown. The i-f amplifier is fixed-tuned and needs no alignment adjustments. The self-tuning of the circuit to 455 kHz is provided by a Clevite TF-01A ceramic filter (M1) which replaces the bypass capacitor across source resistor R4, and by a U. S. Sonics A25H4C ceramic filter (M2) inserted between the two i-f stages.

The 1N295 diode, X1, is a combined second detector and agc rectifier. The volume-control potentiometer (R9) serves the additional function of d-c load resistor for this diode, and the agc voltage is taken from the top of the potentiometer and fed back to the first i-f stage through filter C9-R10. The audio component of the detector output is taken from the potentiometer through coupling capacitor C13 and is presented to the gate of the FET driver (Q4). The class-B coupling transformers are miniature transistor-type units: Input transformer T3 (Allied Radio 54-D-2389, or equivalent) has a 10K primary and 3K center-tapped secondary; output transformer T4 (Allied Radio 54-D-2358, or equivalent) has a 500-ohm center-tapped primary and a 4-ohm secondary.

The receiver delivers audio-output power of approximately 110 milliwatts to the speaker. With a d-c supply of 9 volts, the current drain is 11 ma at zero signal; the drain is 260 ma when a maximum signal is received.

All fixed resistors are 1/2 watt, and electrolytic capacitors C14 and C15 are rated at 25 dcwv. Each FET has a fourth pigtail which is connected to the metal shell internally and must be grounded as shown.

REGENERATIVE BROADCAST RECEIVER

The regenerative broadcast-band receiver circuit shown in Fig. 4-2 will interest experimenters and hobbyists. This circuit, employing one 2N3578 FET, makes a sensitive receiver which is surprisingly selective for a regenerative setup. It picks up local stations with a self-contained ferrite loop antenna, L1, and can bring in weak and distant stations with an outside antenna and ground.

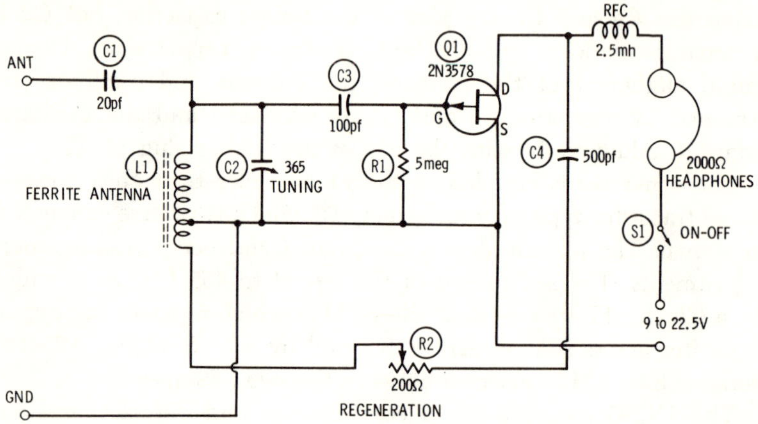

Fig. 4-2. Regenerative broadcast receiver.

The 365-pf variable capacitor, C2, tunes the loop antenna throughout the standard-broadcast band (540-1650 kHz), and the 200-ohm rheostat, R2, provides smooth control of regeneration from zero to full oscillation. Very strong stations come in satisfactorily with no regeneration at all (i.e., with R2 set to its maximum resistance), whereas weak stations are missing unless a good deal of regeneration is used. When the listener is hunting for a very weak station, he can use maximum regeneration, which will produce a whistle when the station carrier is located. He then can reduce the regeneration to the point at which the whistle just disappears.

The circuit operates at battery voltages between 9 volts and 22.5 volts, the higher voltages giving highest sensitivity and loudest signals. With 2000-ohm headphones in the circuit, the current drain is 3 ma at 9 volts, and 4 ma at 22.5 volts.

For increased volume, as for speaker operation, a suitable transistorized audio amplifier may be added to the receiver through a coupling transformer substituted for the headphones.

ALL-WAVE REGENERATIVE RECEIVER

The simplest possible sensitive receiver for all-wave general-coverage and amateur-band phone and c-w reception is the regenerative type. Fig. 4-3A shows the circuit of a receiver of this kind which tunes from 3.5 to 35 MHz with five plug-in coils.

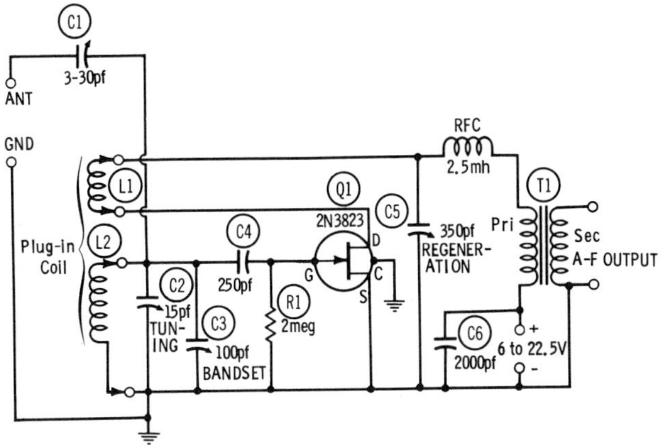

(A) Circuit.

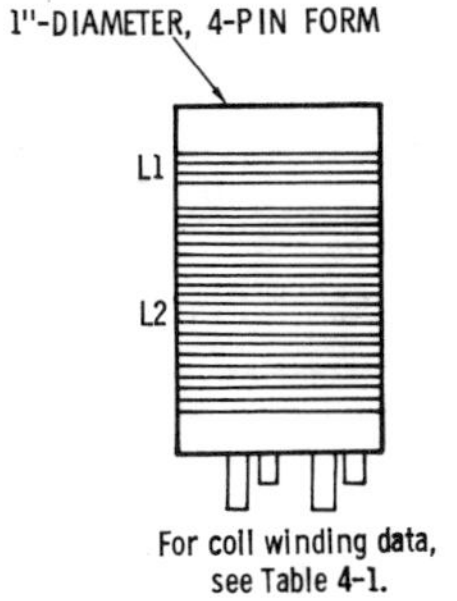

(B) Coil Detail.

Fig. 4-3. All-wave regenerative receiver.

A simple detector with drain tickler feedback is employed. L1 is the tickler coil, and L2 the tuning coil. Both coils are wound on a 1-inch-diameter, 4-pin, plug-in form, as shown in Fig. 4-3B. Variable bypass capacitor C5 is the regeneration control and is a midget 350-pf single-section unit. Audio output is delivered by T1, which may be any convenient transistor- or tube-type interstage transformer with a step-up turns ratio. Either headphones or an a-f amplifier may be operated at the A-F OUTPUT terminals.

The d-c operating voltage may be any available value between 6 and 22.5 volts, the higher values giving best sensitivity and loudest signals. The current drain is 4 ma at 6 vdc, and 8 ma at 22.5 vdc.

Table 4-1 gives coil-winding data. Some experimentation with the number of turns shown in this table may be necessary, since the

Table 4-1. Coil Winding Data for All-Wave Regenerative Receiver*

80-METER BAND (3.5-4 MHz)
L1: 20 turns No. 30 enameled wire, closewound immediately at top of L2 L2: 50 turns No. 24 enameled wire, closewound
40-METER BAND (7-7.3 MHz)
L1: 8 turns No. 30 enameled wire, closewound 1/8 inch from top of L2 L2: 17 turns No. 22 enameled wire, spaced to winding length of 3/4 inch
20-METER BAND (14-14.35 MHz)
L1: 6 turns No. 22 enameled wire, closewound 1/4 inch from top of L2 L2: 10 turns No. 22 enameled wire, spaced to winding length of 3/4 inch
15-METER BAND (21-21.45 MHz)
L1: 4 turns No. 22 enameled wire, closewound 1/4 inch from top of L2 L2: 7 turns No. 22 enameled wire, spaced to winding length of 3/4 inch
10-METER BAND (28-29.7 MHz)
L1: 3 turns No. 22 enameled wire, closewound 1/4 inch from top of L2 L2: 4 turns No. 22 enameled wire, spaced to winding length of 3/4 inch

*All coils wound on 1-inch diameter, 4-pin coil forms (Millen 45004 or equivalent). See Fig. 4-3(B).

FET and circuit capacitances will vary in individual layouts and will alter tuning characteristics. Coils L1 and L2 must be wound in the same direction. The coils may be plugged into a standard tube socket.

If the receiver is used for ham-band reception, C3 will be used to set the receiver to the top of the band (with C2 set to zero capacitance) and bandspread tuning will be afforded by C2. For general-coverage tuning, C2 must be set to minimum (zero) capacitance and the receiver tuned by means of C3. Antenna trimmer C1 must be adjusted for minimum detuning by the outside antenna used.

SINGLE-FREQUENCY, CRYSTAL-CONTROLLED CONVERTER

Fig. 4-4 shows the circuit of a crystal-controlled superhet converter which may be used ahead of an available i-f amplifier/detector/a-f amplifier channel. This arrangement is based on 2N3823 FET's as mixer (Q1) and oscillator (Q2). This arrangement is intended for highly stable single-frequency reception, such as the monitoring of one station in standby service for communication or control purposes.

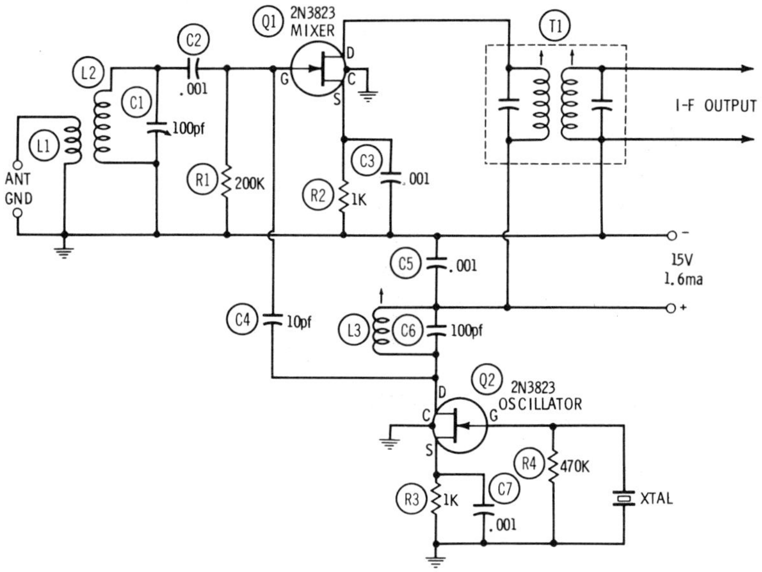

Fig. 4-4. Crystal-controlled converter.

The antenna input coupler is composed of coils L1 and L2, and tuning capacitor C1. The inductance of L2 is selected such that the desired signal frequency (f_s) can be reached by varying C1, and the frequency (f_x) of the crystal (XTAL) is chosen lower than the signal frequency by the amount of the intermediate frequency (f_{if}), i.e., $f_x = f_s - f_{if}$. Similarly, i-f transformer T1 must be tunable to the desired intermediate frequency. The primary coil (L1) of the antenna coupler will contain one-fourth to one-third the number of turns in the secondary coil (L2). The oscillator is tuned to the crystal frequency by adjustment of coil L3.

As a design example, consider that the converter is to operate into a 455-kHz i-f channel and that the received signal will be 5 MHz. (1) The crystal frequency must be 5000 − 455 = 4545 kHz. (2) To tune to this frequency, with C6 = 100 pf, L3 must be adjustable to 12.2 μh, neglecting circuit and FET capacitances. (3) For good leeway of adjustment, let the 5-MHz signal be tuned in with C1 set to half-range, i.e., to 50 pf. This will require that L2 be 20 μh, neglecting circuit and FET capacitances. This value of inductance is obtainable in commercial slug-tuned coils to which a primary (L1) may easily be added by winding insulated wire around the bottom of the coil. Alternatively, a 20-μh coil system may be wound on a 1-inch-diameter form, with L2 = 6 turns of No. 22 enameled wire spaced to a winding length of 1/2 inch, and L1 = 3 turns of No. 22 enameled wire closewound 1/8 inch from the bottom of L2.

For stability, all wiring must be as short, rigid, and direct as practicable. A fourth pigtail of the 2N3823 is internally connected to the metal case of this FET, and for shielding purposes should be grounded as shown. All resistors are 1/2 watt; and for maximum stability, all capacitors should be silvered mica.

At 15 vdc, the total current drain is 1.6 ma.

BEAT-FREQUENCY OSCILLATOR

In a standard superhet receiver, a beat-frequency oscillator (bfo) is needed for c-w reception. The usual procedure is to operate the bfo at the intermediate frequency and to couple its output into the second detector. A slight detuning of the bfo then will cause a beat note with the i-f carrier and will make the c-w signal audible. Fig. 4-5 shows the circuit of a bfo, using a U183 FET, which may be used with a receiver having no bfo of its own.

This Colpitts-type circuit operates at 455 kHz, the oscillation frequency being determined by 2.5-mh inductor L1 (which may be a conventional r-f choke of that inductance) and the series-parallel arrangement of capacitors C1, C2, and C3. The frequency is set initially, with C3 set to midcapacitance, by adjustment of the tuning slug in coil L1 (Miller 21A223RBI, or equivalent). Subsequent tuning of variable capacitor C3 adjusts the pitch of the beat note to suit the operator's ear.

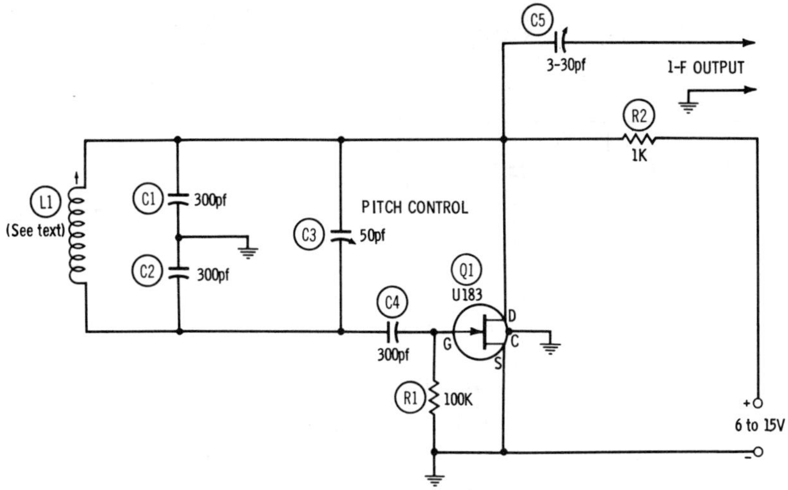

Fig. 4-5. Beat-frequency oscillator.

At 6 vdc, the current drain is 2 ma and the no-load i-f output is 5 volts peak to peak; at 9 vdc, the current drain is 2.1 ma and the i-f output 7 volts peak to peak; and at 15 vdc, the current drain is 2.2 ma and the i-f output 8.5 volts peak to peak. The bfo output is coupled to the second detector in the receiver through a 3-30–pf trimmer capacitor, C5. This capacitor must be set for optimum signal for a smooth beat note without receiver blocking.

Resistors R1 and R2 are 1/2 watt, and capacitors C1 and C2 are silvered mica. For stability, all wiring must be kept short, rigid, and direct, and the fourth pigtail of the U183 (which is internally connected to the metal case of this FET) should be grounded as shown.

The bfo may be adapted to a higher intermediate frequency by using a lower-inductance coil in place of the 2.5-mh unit shown in Fig. 4-5.

Q-MULTIPLIER

The Q-multiplier is a handy device for boosting the selectivity of a receiver in either phone or c-w reception. This device is especially valuable when the receiver has no crystal filter. Regeneration in the Q-multiplier sharpens the response of the i-f channel so as to peak a desired signal or reject an undesired one. Fig. 4-6 shows the circuit

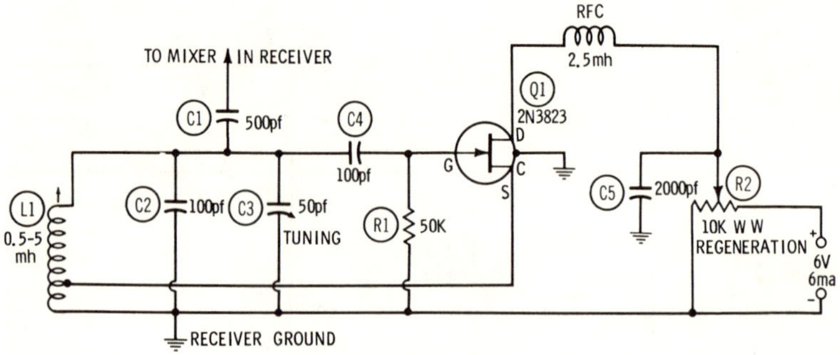

Fig. 4-6. Q-multiplier.

of a simple Q-multiplier employing a 2N3823 FET and designed for receivers having a 455-kHz i-f channel. The output of this instrument is coupled, through capacitor C1, to the plate, collector, or drain in the mixer stage of the receiver (depending upon whether the receiver uses tubes, bipolar transistors, or field-effect transistors).

The Q-multiplier shown here is a 455-kHz Hartley oscillator with controlled regeneration. The operating frequency is determined by coil L1 (Miller 6323, or equivalent) which is adjustable between 0.5 and 5 mh, and capacitors C2 and C3. Initially, the instrument is set to 455 kHz, with C3 set to midcapacitance, by adjustment of the tuning slug in the coil. Capacitor C3 then is used subsequently to tune the multiplier *sharply* to the receiver intermediate frequency.

The amount of regeneration, and thus the selectivity of the receiver, is controlled by varying the FET d-c voltage with the 10,000-ohm potentiometer, R2. Two modes of operation are obtained: with a certain amount of regeneration, the Q-multiplier acts as a notch filter, eliminating a signal to which the receiver is tuned; with more regeneration, the i-f response is peaked, tuning the receiver sharply to one desired signal coming in on the set.

The shortest possible leads must be used between Q-multiplier and receiver. Also, all wiring inside the Q-multiplier must be short, rigid, and direct. Resistor R1 is 1/2 watt, and capacitors C1, C2, and C4 are silvered mica. The instrument should be shielded by enclosing it in a metal box, and for additional shielding the fourth pigtail of the 2N3823 (being connected internally to the metal case of this FET) must be grounded as shown.

At 6 vdc, the total current drain is 6 ma. The d-c operating power for the Q-multiplier may be obtained from a self-contained battery or from the receiver power supply.

"SELECTOJECT"

The "Selectoject," like the Q-multiplier and the crystal filter, permits either the selection or rejection of a signal but on the basis of the latter's audio frequency. The "Selectoject" is connected between a-f stages of a receiver, or between second detector and audio channel.

Fig. 4-7A shows the circuit of a transistorized "Selectoject" employing four TIS14 FET's. Fig. 4-7B shows performance of the circuit. In this arrangement, which is essentially a sharply tuned a-f amplifier, a dual rheostat (R4) serves as the tuning control. One rotation of this control tunes the "Selectoject" from 100 Hz to 10 kHz. Potentiometer R9 serves as the selectivity control (for adjusting the sharpness of response to suit individual conditions) and dpdt switch S1 serves as the function switch (for selecting either the pass or reject function of the circuit).

With the function switch in its PASS position, the response at the frequency to which the "Selectoject" is tuned is shown by the solid curve in Fig. 4-7B. With this switch in its REJECT position, the response is shown by the dotted curve. The height of the pass curve or the depth of the reject curve, and the width of either curve, are controlled by the setting of potentiometer R9, highest selectivity in either case resulting when R9 is set for maximum gain in FET Q4.

The select/reject response of the circuit results from the fact that signal through the Q4 stage (from input to output of the entire circuit) is negative when the function switch is thrown to REJECT and therefore cancels the tuned-in signal, whereas this feedback (this time, from the source of FET Q3 to the input of the circuit) is positive when the function switch is thrown to PASS, and therefore boosts the tuned-in signal. If potentiometer R9 is advanced far enough when the function switch is set to PASS, the circuit will oscillate, delivering a sine-wave signal at the frequency to which tuning control R4 is set.

The resistance of R2 must be selected equal to the resistance of R3 within 1 percent. Similarly, the resistance of R5 must be selected equal to the resistance of R6 within 1 percent. Also, the capacitance of C2 must equal the capacitance of C3 within 1 percent. All fixed resistors are 1/2 watt.

For stability, all wiring must be kept as short, rigid, and direct as practicable. A metal case, to which the circuit is grounded, will

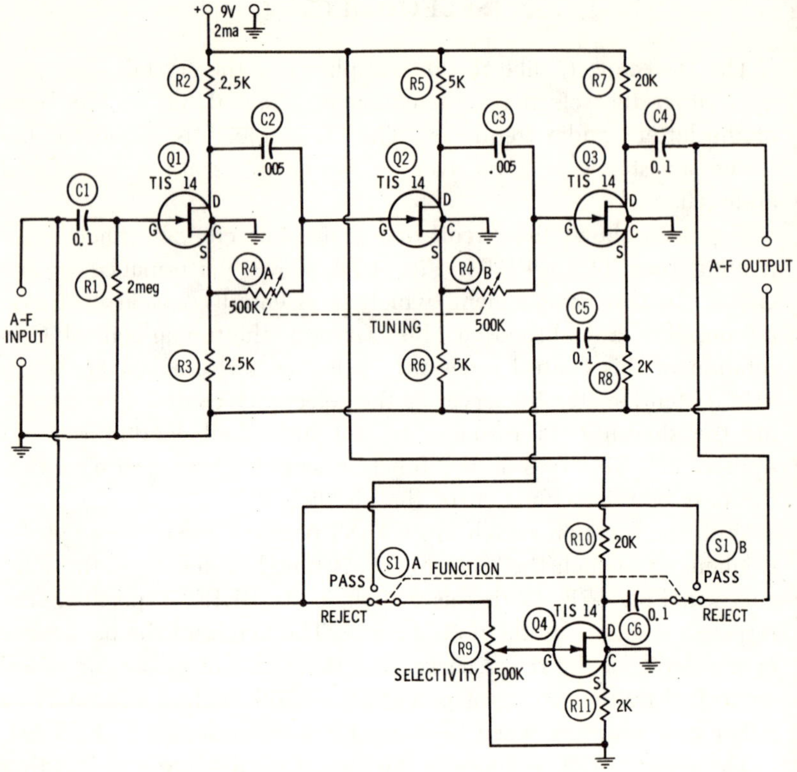

(A) Circuit.

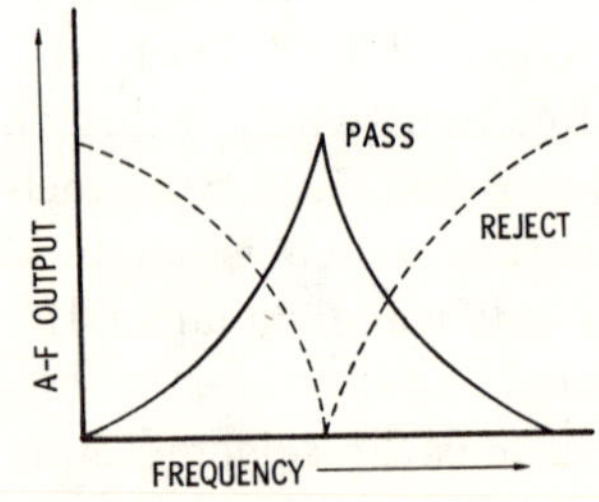

(B) Performance.

Fig. 4-7. "Selectoject."

shield the "Selectoject" from external fields, and for additional shielding the fourth pigtail of the TIS14 (which is internally connected to the metal case of this FET) must be grounded as shown. Total current drain is 2 ma at 9 vdc.

AUDIO SQUELCH AMPLIFIER

Fig. 4-8 shows the circuit of a simple squelch amplifier which can hold the audio channel of a receiver silent until the receiver signal reaches a predetermined amplitude. This single amplifier stage may be inserted at any convenient point near the front end of the audio channel.

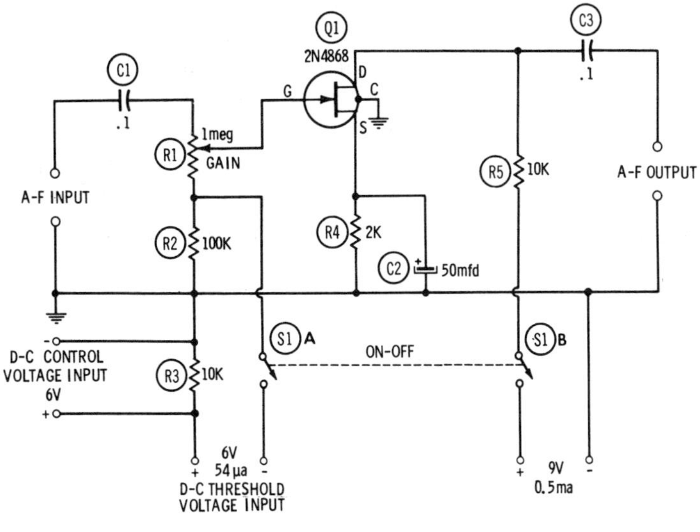

Fig. 4-8. Audio squelch amplifier.

The circuit is a simple r-c–coupled amplifier, based on a 2N4868 FET (Q1), which is biased to cutoff by a d-c gate voltage applied to the THRESHOLD terminals. A d-c voltage, applied to the D-C CONTROL VOLTAGE terminals, bucks this bias and activates the amplifier whenever it exceeds the threshold cutoff voltage. The control voltage may be derived from the i-f amplifier (by rectification) or from the 2nd detector of the receiver. When the amplifier is in operation, it has a voltage gain of 10, with a maximum undistorted output of 1 volt rms. This means that the a-f input voltage at the gate of the 2N4868 must not exceed 0.1 volts rms for any setting of gain-control potentiometer R1.

Main current drain is 0.5 ma at 9 vdc. Virtually all of the current taken from the 6-volt threshold source flows through resistor R2 and is 54 microamperes. The d-c control voltage appears across resistor R3, and this creates a current of 0.6 ma through that resistor. While separate d-c voltage sources are shown for the main supply and the threshold, these voltages may be obtained from appropriate points in the receiver circuit (as along a voltage divider in the power supply.)

If a particular receiver circuit gives voltage of the polarity opposite to that shown in Fig. 4-8, use a P-channel FET (such as 2N3578), change the values of R4 and R5 as required, and reverse capacitor C2.

Lead dress may be important in individual installations, so all leads must be kept as short and direct as practicable. Route them so as to avoid stray coupling in the receiver. A fourth pigtail of the 2N4868 is internally connected to the metal case of this FET and must be grounded as shown, for shielding. Resistors R2, R3, R4, and R5 are 1/2 watt. Capacitors C1 and C3 are 100-volt plastic units, and C2 is a 25-dcwv electrolytic.

5

Transmitter and Transmitter-Accessory Circuits

The special features of the FET which recommend this transistor for use in oscillators (see Chapter 3) also enable it to perform efficiently in low-powered transmitter circuits. In fact, many tube-type oscillator and r-f amplifier circuits for transmitters need no change other than voltage reduction if the tubes are replaced with suitable FET's. This means that standard coils, tuning capacitors, and chokes may be used, and also that no loss results in capacitance-type interstage coupling in transmitters and exciters. Furthermore, the high input impedance of the FET tends to preserve the high Q of the transmitter tank circuit.

Low-powered circuits offering high overall efficiency and relative simplicity are possible in this area with FET's. The transmitter circuits described in this chapter are of especial interest to the radio experimenter, novice, and ham. But they will also suggest applications to communications professionals. In addition to actual transmitters, some accessory devices for use in transmitting stations are described. Additional devices may be found in Chapter 7.

Operation of transmitters is regulated by the Federal Communications Commission (FCC). Be sure you are complying with their rules and procedures before you attempt to put a signal "on the air."

SINGLE-STAGE CRYSTAL-CONTROLLED TRANSMITTER

Fig. 5-1 shows the circuit of a simple crystal-controlled, oscillator type c-w transmitter with a low-impedance output. This single-stage unit, based on a 2N3823 FET, covers the amateur bands from 10 to 80 meters and is capable of 56 milliwatts d-c power input.

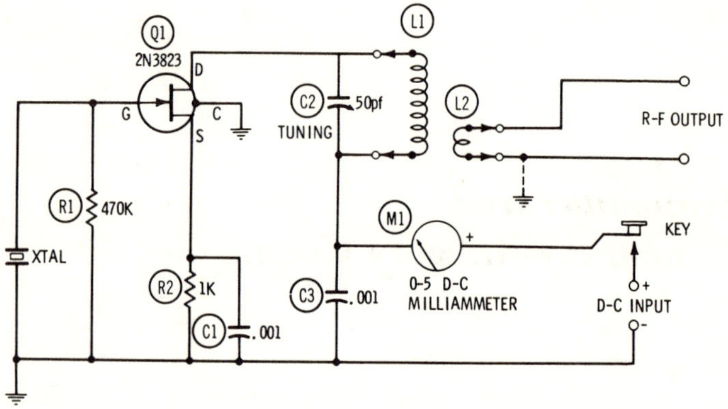

Fig. 5-1. Single-stage 56-mw crystal-controlled transmitter.

The circuit is that of the basic drain-tuned crystal oscillator. The tuned tank circuit comprises 50-pf variable capacitor C2 and tank coil L1. Each plug-in "coil" consists of a tank coil (L1) and link-coupling coil (L2) as a unit (Millen low-powered, single-ended 40,000 series, or equivalents), and separate plug-ins are available for the 10-, 20-, 40-, and 80-meter ham bands. The 20-meter unit can be used also for the 21-meter band. Special coils wound to suit individual demands will cover other frequencies.

The low-impedance output may feed a coaxial antenna feeder directly, or may operate into a suitable antenna coupler. In portable emergency applications, a whip antenna may be connected directly to the top of coil L1.

With the key depressed, the circuit is tuned, without antenna or other load, by adjusting C2 for drain-current dip, as indicated by 0-5 d-c milliammeter M1. The antenna then is connected and any adjustments made to the antenna coupler (if one is used) to "load" the drain current up to the highest value at which the oscillator starts readily each time the key is closed. Drain current is 1.40 ma at 6 vdc, 1.41 ma at 9 vdc, 2.0 ma at 15 vdc, and 2.5 ma at 22.5 vdc. With an active crystal, the oscillator can be made to act as a doubler, i.e., tank L1-C2 may be tuned to twice the crystal frequency. The r-f output will be lower, however, than in straight-through operation.

In addition to serving as a single-stage transmitter, this circuit may also be used as the crystal oscillator in a multistage transmitter. Either inductive or capacitive output coupling may be used in the latter application.

All wiring must be kept as short and rigid as practicable. And the fourth pigtail of the 2N3823 (being internally connected to the metal case of this FET) must be grounded as shown. Resistors R1 and R2 are 1/2 watt; capacitors C1 and C3 are mica.

SINGLE-STAGE 0.8-WATT CRYSTAL-CONTROLLED TRANSMITTER

The U222 FET will provide a maximum of 1 watt r-f output at frequencies as high as 200 MHz. This capability makes the U222 especially attractive for transmitter use.

Fig. 5-2 shows the circuit of a single-stage crystal-controlled c-w transmitter using the U222. This circuit is similar to the one de-

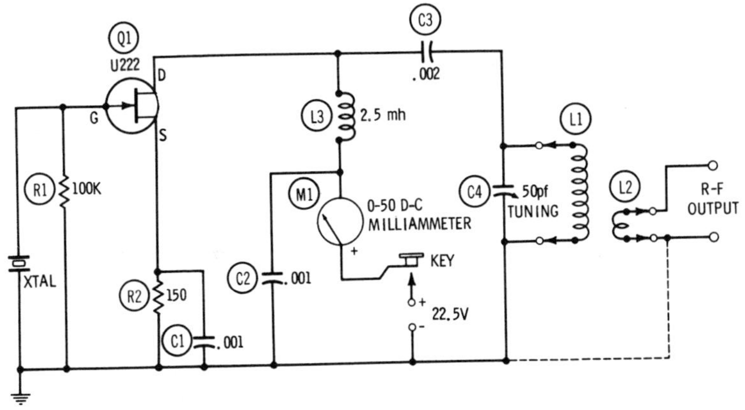

Fig. 5-2. Single-stage 0.8-watt crystal-controlled transmitter.

scribed in the preceding section, except that the U222 circuit employs shunt feed of the d-c supply through a 2.5-mh r-f choke (L3). This allows the bottom of the tank circuit, L1-C4, to be grounded. Plug-in coil sets (L1-L2) permit the transmitter to be tuned to any frequency for which a crystal is available. Each plug-in set consists of a tank coil (L1) and link-coupling coil (L2) assembled as a unit. The Millen 40,000 series of low-powered single-ended coils provides sets for the 10-, 20-, 40-, and 80-meter ham bands (the 20-meter set can be used also for the 21-meter band). For nonham frequencies, other L1-L2 coils may be wound to suit individual demands.

The circuit is tuned, with the key depressed and with no antenna or other load, by adjusting C4 for drain-current dip, as indicated by 0-50 d-c milliammeter M1. Then, the antenna is connected and any adjustments made to the antenna coupler (if one is used) to "load" the drain current up to 35 ma minimum. At this point, the d-c power input is 0.8 watt. The low-impedance output may be connected directly to a coaxial antenna feeder or may operate into a suitable antenna coupler.

Like the preceding single-stage transmitter (Fig. 5-1), this one may serve either as a complete c-w transmitter or as the oscillator of a multistage transmitter.

Resistors R1 and R2 are 1/2 watt, and capacitors C1, C2, and C3 are mica. All wiring must be short and rigid, for stability. The gate electrode of the U222 is internally connected to the metal case of this FET which accordingly must be kept out of contact with other components.

TWO-STAGE TRANSMITTER

Fig. 5-3 shows the circuit of a crystal-controlled oscillator-amplifier type of transmitter, employing single-ended stages. A 2N3823 FET is used in the oscillator (Q1) and a U222 FET in the r-f amplifier (Q2). The d-c power input of the amplifier is 0.8 watt.

The oscillator is tuned by tank circuit L1-C2. Here, L1 is a plug-in coil chosen to resonate with C2 at the crystal frequency. The amplifier is tuned by tank circuit L2-C6, and here a plug-in coil set (L2-L3) is used: L2 resonates with C6 at the crystal frequency, and L3 is a link-coupling coil. For ham bands, the oscillator and amplifier coils both may be commercial low-powered units, such as the Millen 40,000 series of single-ended units. For L1, the main

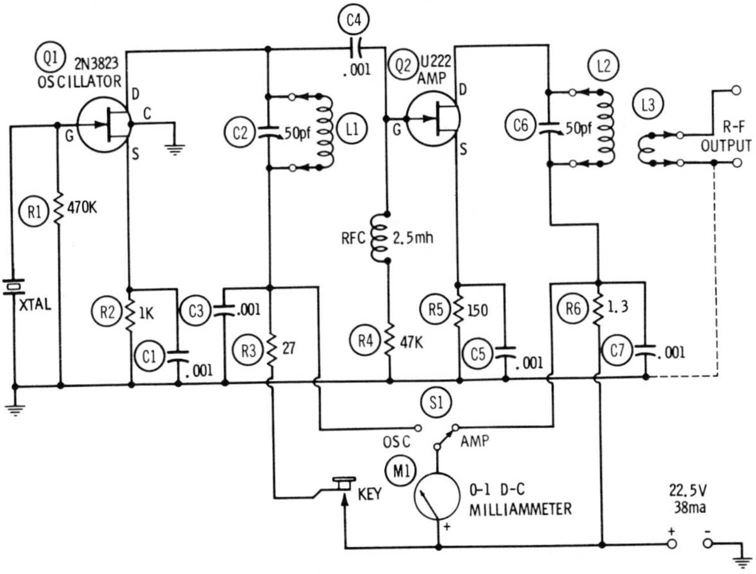

Fig. 5-3. Two-stage transmitter.

coil is used without the link winding. The low-impedance R-F OUTPUT terminals of the transmitter can feed a coaxial antenna feeder directly or can operate into a suitable coupler for a higher-frequency antenna.

A single 0-1 d-c milliammeter (M1) is used for tune-up purposes. When meter switch S1 is in its OSC position, the meter is switched across 27-ohm resistor R3, the scale is converted to 0-3 ma, and the meter reads oscillator current; when S1 is in its AMP position, the meter is switched across 1.3-ohm shunt resistor R6, the scale is converted to 0-43 ma, and the meter reads amplifier current. (Both shunt resistors are stock values). The transmitter is tuned to the crystal frequency, in the conventional manner, by adjusting first the oscillator capacitor (C2) and then the amplifier capacitor (C6) for drain-current dip. Common drain-current values are 2.5 ma for the oscillator and 35 ma for the amplifier.

All wiring must be kept short, direct, and rigid as practicable. Overall shielding, such as complete enclosure in a metal box, is advantageous but not mandatory. However, the fourth pigtail of the 2N3823 (being internally connected to the metal case of this FET) must be grounded as shown. The gate electrode of the U222 is in-

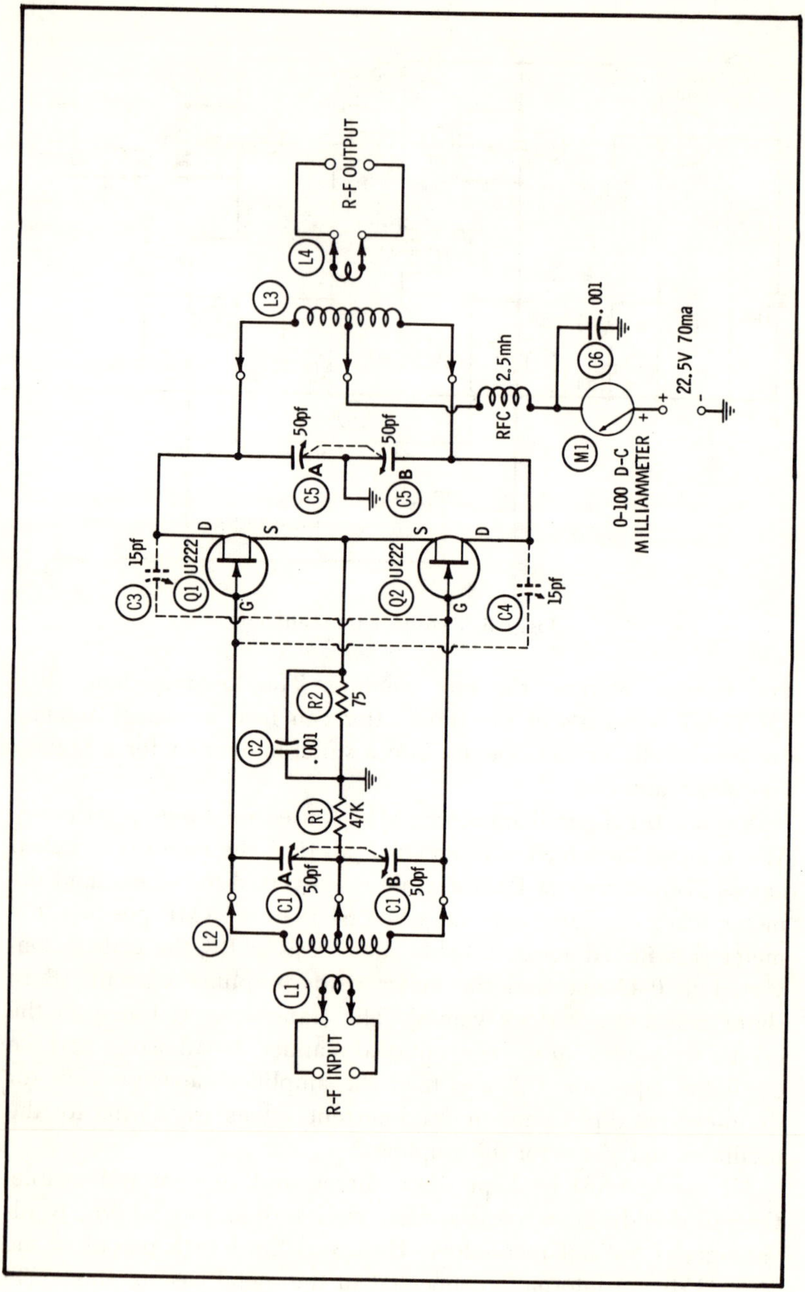

Fig. 5-4. Push-pull final amplifier.

ternally connected to the metal case of that FET, so the U222 must be kept clear of all contact with other components and wiring. All fixed resistors are 1/2 watt, and all fixed capacitors are mica.

PUSH-PULL FINAL AMPLIFIER

A push-pull final amplifier capable of approximately 1.6 watts d-c power input is shown in Fig. 5-4. This amplifier may be driven by the two-stage circuit shown in Fig. 5-3 or by the crystal oscillator shown in Fig. 5-2. Two U222 FET'S are used.

The amplifier employs link-coupled input and output (low-impedance coupling coils L1 and L4). For ham bands, plug-in coils (such as the Millen 40,000 series, low-powered, center-tapped type) may be used both for the input (L1-L2) and output (L3-L4) tanks. Both tanks are tuned to the incoming frequency.

In many instances, the amplifier will self-oscillate and will need to be neutralized. The 15-pf variable capacitors (C3, C4) are indicated for this purpose, but will not be needed in every layout. The circuit should be tried first without them, noting if self-oscillation occurs at any setting of tuning capacitors C1A-C1B and C5A-C5B. Netutralizing is accomplished in a conventional manner.

All wiring must be kept short, direct, and rigid; and care should be taken to avoid contact between the FET's or between them and other components, since the gate electrode is internally connected to the metal case of the U222. Resistors R1 and R2 are 1 watt, and capacitors C2 and C6 are mica.

Current drain is approximately 70 ma at 22.5 volts, but may be higher with individual FET's.

100-MHz GROUNDED-GATE AMPLIFIER

The grounded-gate FET amplifier is the counterpart of the grounded-grid tube amplifier. Neither of these amplifiers requires neutralization; hence, both are usable at very high frequencies in circuits which are simpler than they otherwise might be. The grounded-gate (and grounded-grid) amplifier has the additional advantage of somewhat higher output than that of the grounded source, grounded cathode, or grounded emitter under similar operating conditions, since most of the excitation power also appears in the amplifier output.

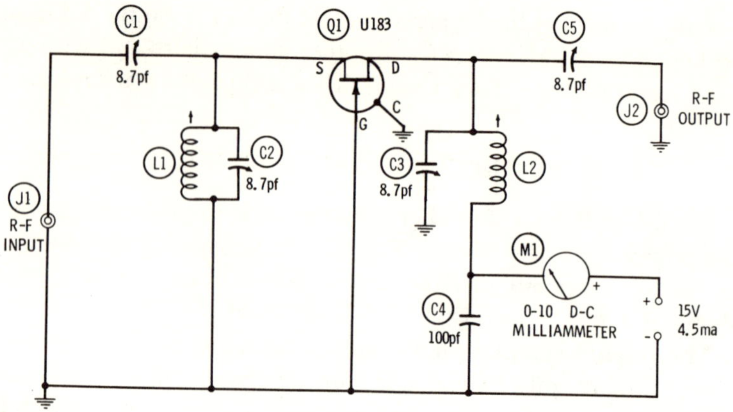

Fig. 5-5. 100-MHz grounded-gate amplifier.

Fig. 5-5 shows the circuit of a grounded-gate r-f amplifier suitable for 100-MHz use. Employing a U183 FET, this amplifier has 50-ohm input and output (available through coaxial jacks J1 and J2, respectively).

In this arrangement, the input (source-to-gate) circuit is tuned by means of tank L1-C2, and the output (drain-to-gate) circuit by means of tank L2-C3. Variable capacitors C1 and C5 are used for coupling and impedance matching. Each of the variable capacitors (C1 to C5) has a maximum capacitance of 8.7 pf (E. F. Johnson 160-104 midget, or equivalent) and each of the tank coils (L1, L2) is a slug-tuned 0.184-0.275-μh unit (Miller 40A227CBI, or equivalent). The coil slugs are set to tune the tanks to 100 MHz with C2 and C3 at half-capacitance.

Current drain is 4.5 ma at 15 vdc; this corresponds to a d-c power input of 67.5 mw. The r-f input (driving) power is approximately 4 mw.

All wiring must be kept as short, direct, and rigid as practicable; and the fourth pigtail of the U183, being internally connected to the metal case of this FET, must be grounded as shown. Fixed capacitor C4 must be silvered mica.

SINGLE-ENDED FREQUENCY MULTIPLIER

Fig. 5-6 shows the circuit of a single-ended frequency-multiplier stage. Although this circuit gives best efficiency as a frequency

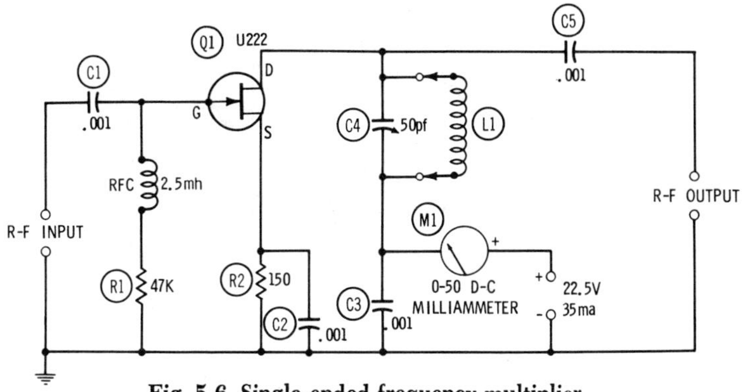

Fig. 5-6. Single-ended frequency multiplier.

doubler, it may also be used as a tripler. Operation is entirely conventional: the output tank (L1-C4) simply is tuned to twice or to three times the frequency of the r-f input, as desired.

The circuit, which employs a single U222 FET, may be driven by either of the crystal oscillators described earlier (Fig. 5-1 and 5-2), but substantially higher output as a doubler or as a tripler will be obtained when the oscillator of Fig. 5-2 is used as the driver.

Inductance L1 is selected to resonate with capacitance C4 at the desired multiple of the input-signal frequency. For the ham bands, plug-in coils may be any of the commercial low-powered units, such as the Millen 40,000 series. Either capacitive output coupling may be used, as shown in Fig. 5-6, into a high-impedance load; or low-impedance inductive coupling may be obtained with the link coil supplied as part of the commercial plug-in coils (L1).

The d-c power input is approximately 0.8 watt (35 ma at 22.5 vdc). Drain-current dip, as indicated by 0-50 d-c milliammeter M1, indicates correct tuning of the multiplier.

All wiring must be kept short, direct, and rigid; and since the gate electrode of the U222 is internally connected to the metal shell of this FET, the latter must be kept out of contact with other components and wiring. All resistors are 1/2 watt, and all fixed capacitors are mica.

PUSH-PUSH DOUBLER

The push-push doubler, using two tubes or transistors, gives approximately twice the r-f power output of the single-ended doubler.

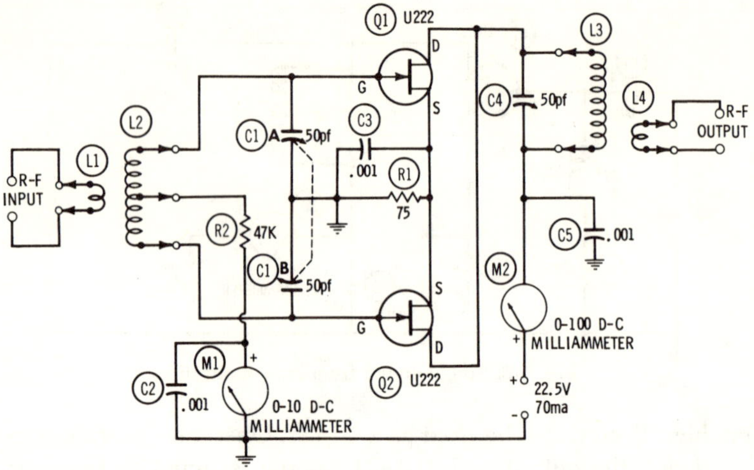

Fig. 5-7. Push-push doubler.

Fig. 5-7 shows the circuit of a push-push doubler employing two U222 FET's.

Note that, in the conventional manner, the gates are connected in push-pull and the drains in parallel. Because of this arrangement, a center-tapped input tank coil (L2) and single-ended output tank coil (L3) are required. (Both tanks are link coupled, the input through L1 and the output through L4). The inductance of L2 is selected to resonate with dual tuning capacitor C1A-C1B at the input frequency, and the inductance of L3 to resonate with tuning capacitor C4 at twice the input frequency. Input- and output-coil sets may be low-powered commercial plug-in units, such as the Millen 40,000 series.

Tuning procedure is conventional: Tune the input tank to the input frequency by adjusting C1A-C1B for peak deflection of milliammeter M1, then adjust the output tank for drain-current dip at twice the input frequency, as indicated by milliammeter M2. The d-c power input is approximately 1.6 watt (70-ma drain current at 22.5 vdc).

All wiring must be as short, direct, and rigid as practicable. Resistors R1 and R2 are 1 watt, and all fixed capacitors are mica. The U222's must be kept clear of contact with each other and with wiring and other components, since the gate electrode is internally connected to the metal case of this FET.

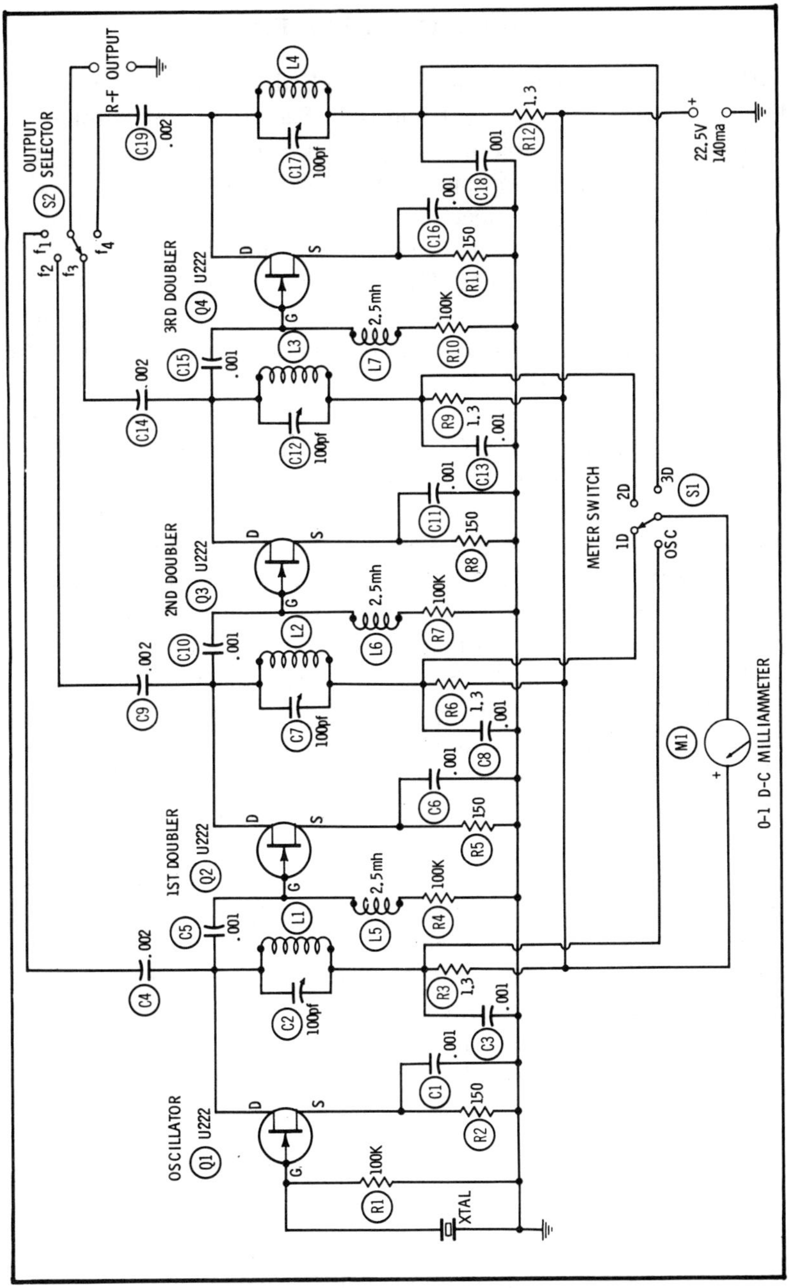

Fig. 5-8. Four-band exciter.

FOUR-BAND EXCITER

Fig. 5-8 shows the circuit of a standard four-band exciter for ham radio. This arrangement consists of a crystal oscillator (Q1) followed by three doublers (Q2, Q3, Q4), each stage containing a single U222 FET. The output (f1) of the oscillator or that of the doublers (f2, f3, f4) is selected, as desired, by means of the single-pole, 4-position rotary switch, S2. All r-f outputs are capacitance coupled (through C4, C9, C14, and C19, respectively). In a popular application of this type of exciter, the crystal oscillator operates in the 80-meter band, and the doublers in the 40-, 20-, and 10-meter bands, respectively.

The various stages are tuned by means of separate tank capacitors: C2 for the oscillator; C7, first doubler; C12, second doubler; and C17, third doubler. Each stage is tuned, in the conventional manner, starting with the oscillator and progressing through the exciter, for drain-current dip. For economy in this tuning, a single 0-1 d-c milliammeter (M1) is switched (by means of S1) across a shunt resistor (R3, R6, R9, R12) in any stage being adjusted. These resistances are low enough that they introduce only negligible loss in the exciter, but multiply the meter to approximately 0-42 ma full scale. The scheme is much simpler than would be required if the lines had to be broken to insert the meter and closed after completion of the measurement. The 1.3-ohm resistors are stock 5 percent units. Where economy is unimportant, however, a separate 0-50 d-c milliammeter may be installed in each stage.

The d-c power input is approximately 0.8 watt per stage (35 ma at 22.5 vdc). Total current drain for the exciter is 140 ma.

All wiring must be kept as short, direct, and rigid as practicable. All resistors are 1/2 watt, and all fixed capacitors are mica. The U222's must be kept clear of contact with each other and with wiring and other components, since the gate electrode is internally connected to the metal case of this FET.

BALANCED MODULATOR

For single-sideband operation, the FET makes possible a triode-type balanced modulator which is compact and low powered. Fig. 5-9 shows such a balanced-modulator circuit employing two matched 2N4340 FET's. Note that the a-f (modulating) signal is applied to the gates in push-pull through transformer T1 and that the r-f (carrier) signal is applied to the sources in parallel through capacitor C3.

Cancellation of the carrier in the output circuit (only the two sidebands which result from the modulation remaining) results from

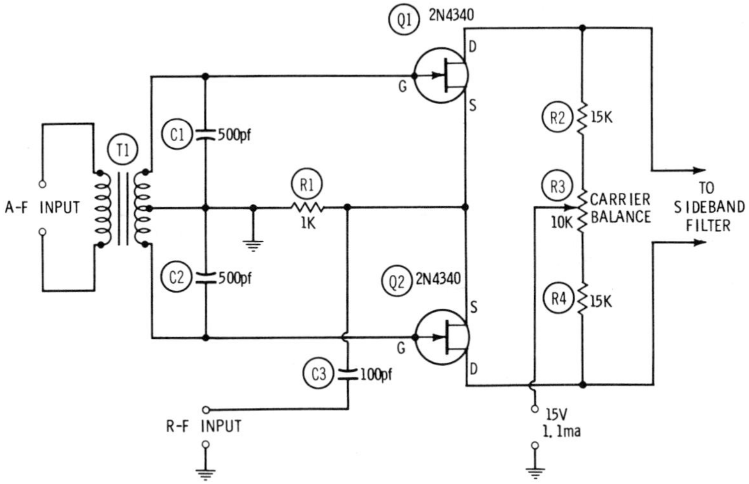

Fig. 5-9. Balanced modulator.

the symmetry of the circuit. To ensure this symmetry, there must be close correspondence between the two halves of the circuit. This means that the secondary winding of transformer T1 must be accurately center tapped, the two FET's must be matched, C1 must match C2, and R2 must match R4. Any slight unbalance due to these components or to circuit strays may be corrected by adjustment of the CARRIER BALANCE potentiometer, R3. Total current drain is 1.1 ma at 15 vdc.

All wiring must be kept as short, direct, and rigid as practicable, and the circuit should be enclosed within a grounded shield box. All fixed resistors are 1/2 watt, and all fixed capacitors are silvered mica. The 2N4340's must be kept clear of contact with each other and with wiring and other components, since the gate electrode is internally connected to the metal case of this FET.

The terminal arrangement of the FET's can be checked against the drawings shown in Appendix A. Manufacturers may be obtained from Appendix B, and the addresses are shown in Appendix C.

C-W MONITOR

Fig. 5-10 shows the circuit of a simple monitor for c-w telegraphy. Requiring no connection to the transmitter and containing its own power supply, this device may be placed conveniently on the operating desk. It will give a reliable aural check of signal quality, as well as making the keying audible.

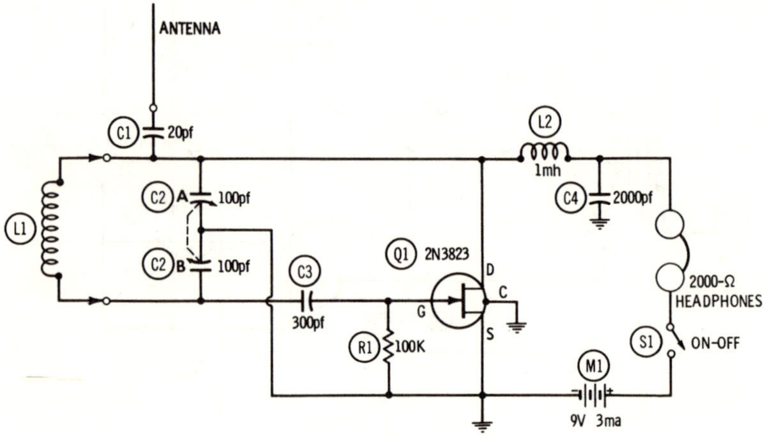

Fig. 5-10. C-w monitor.

The circuit is that of a Colpitts-type oscillating detector employing a single 2N3823 FET. The signal is tuned in by the dual 100-pf variable capacitor C2A-C2B. The antenna may be any convenient whip; short, stiff, vertical wire; or flexible, insulated, scatter lead which will pick up enough signal for a comfortable headphone level. The tuning capacitor may be detuned above or below the signal for the most pleasing beat tone. Total current drain is 3 ma at 9 vdc.

Inductance L1 is chosen to resonate with one-half C2A-C2B (i.e., with 50 pf) at the signal frequency. For ham-band operation, a set of low-powered commercial plug-in coils, such as the Millen 40,000 series, may be used to cover the 3.5- to 30-MHz spectrum.

The volume delivered by a pair of high-impedance magnetic headphones will be sufficient in most environments, especially with medium- and high-power transmitters. However, a transistorized audio amplifier may be added to the monitor if speaker operation or louder headphone signals are desired.

All wiring must be kept as short, direct, and rigid as practicable, and the monitor circuit should be enclosed in a shield box to which the circuit is grounded. For further shielding, the fourth pigtail of the 2N3823 (which is internally connected to the metal case of this FET) must be grounded as shown. Resistor R1 is 1/2 watt, and capacitors C1, C3, and C4 are mica. The terminal arrangement for the FET is shown in Appendix A.

MODULATING THE FET TRANSMITTERS

The FET transmitters shown in this chapter are depicted for c-w telegraph operation. They may be amplitude modulated for telephony. however, by short-circuiting the telegraph key and connecting a suitable a-f speech amplifier/modulator unit in series with the transmitter d-c supply, as shown in Fig. 5-11.

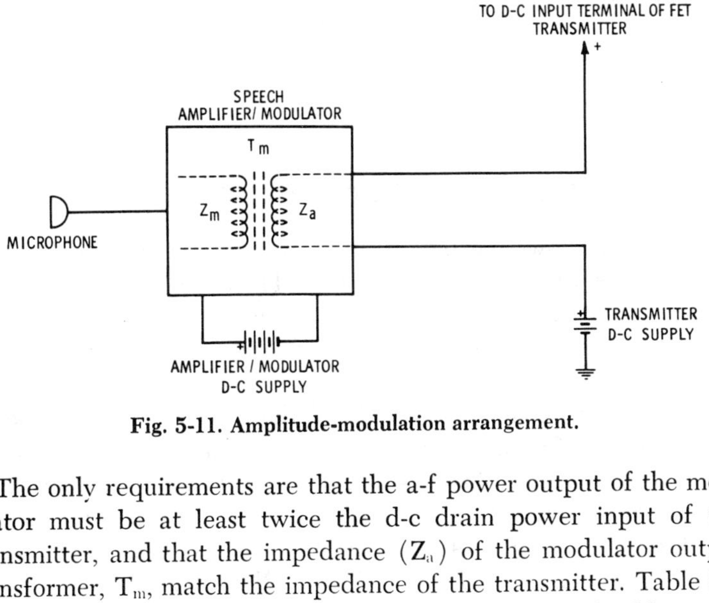

Fig. 5-11. Amplitude-modulation arrangement.

The only requirements are that the a-f power output of the modulator must be at least twice the d-c drain power input of the transmitter, and that the impedance (Z_a) of the modulator output transformer, T_m, match the impedance of the transmitter. Table 5-1 shows the values of impedance (for each of the applicable transmitters in this chapter) which transformer T_m in the modulator must match. It also shows the a-f output power needed to modulate each of the transmitters 100 percent. The primary impedance (Z_m) of the transformer matches the impedance of the transistor(s) in the modulator.

Table 5-1. A-M Modulator Data

Transmitter Circuit	Transmitter D-c Voltage	A-f Power (Milliwatts)	Transmitter Impedance (Ohms)
Fig. 5-1*	6.0	16.8	4290
	9.0	25.4	6360
	15.0	60.0	7500
	22.5	56.2	9000
Fig. 5-2*	22.5	1600	645
Fig. 5-3	22.5	1600	645
Fig. 5-4	22.5	3200	32.2
Fig. 5-5	15.0	135	3300
Fig. 5-6	22.5	1600	645
Fig. 5-7	22.5	3200	32.2

*Modulated oscillators are prohibited in some services.

Note that the modulation must be applied to the final r-f stage in the transmitter circuit given in Fig. 5-3.

SINGLE-CONTROL ("TNT") TRANSMITTER

Fig. 5-12 shows the circuit of a c-w transmitter. This is a variation of the tuned-plate, tuned-grid (TPTG) oscillator which uses tuning only in the output circuit. An untuned coil is used in the input circuit. This accounts for the descriptive term *TNT* ("tuned, nontuned").

In the arrangement shown here, the drain circuit is tuned by means of 50-pf variable capacitor C2 and a plug-in coil. Each plug-in coil set (Millen 43,000 Series, or equivalent), consists of an untapped main coil (L4) and end link (L5). Three sets provide coverage of the 3.5-, 7-, and 14-MHz amateur bands. Corresponding untuned gate coils (L1, L2, and L3) for these bands are selected by means of single-pole, three-position switch S1. The gate coils are preadjusted by means of their screwdriver-operated tuning slugs. (L1, 3.5-MHz band, Miller No. 40A475CBI. L2, 7-MHz band, Miller No. 40A155CBI. L3, 14-MHz band, Miller No. 40A336CBI).

The U222 FET (Q1) draws 35 ma at 22.5 volts dc, a d-c input corresponding approximately to 0.8 watt. Lower d-c voltages will

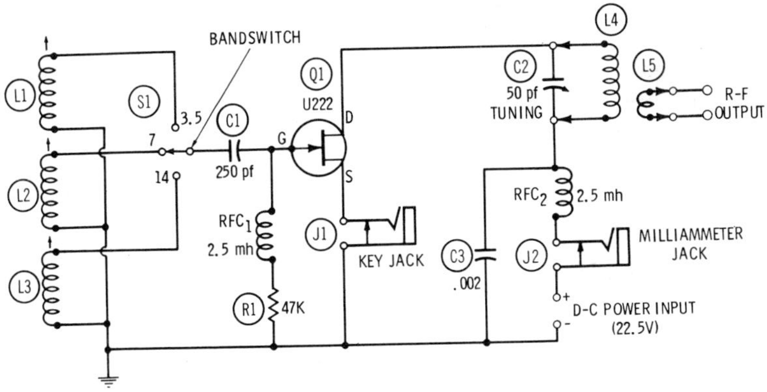

Fig. 5-12. Single-control ("TNT") transmitter.

produce correspondingly lower power input. A 0 to 50 d-c milliammeter plugged into jack J2 allows drain current to be read at intervals, or it can be monitored continuously.

In operation, the transmitter may be tuned throughout each band simply by adjusting variable capacitor C2. There is sufficient distributed capacitance and circuit capacitance across the gate coil (L1, L2, L3) in most layouts that this coil, in true TNT fashion, needs no separate tuning. However, the transmitter output will be *maximum* around the one frequency at which the gate coil tends to self-resonate, and will fall off somewhat at higher and lower frequencies. The optimum gate-coil inductance which will give best output throughout the band may be selected by adjusting the gate-coil slug, varying capacitor C2, and noting the relative output as indicated by an r-f vtvm connected to the R-F OUTPUT terminals, or by a 2-volt pilot lamp (with a 1-turn pickup coil) coupled to tank coil L4. In most instances, the gate coil will be adjusted for peak output at the center of the band. When maximum efficiency is desired at some extreme frequency in any band, the appropriate gate coil may be retrimmed for maximum output at that frequency.

The gate coils must be mounted at right angles to each other and separated as widely as practicable, to prevent power absorption by the unused two. For maximum isolation, L1, L2, and L3 may be enclosed in separate shield cans.

Closed-circuit jacks are used both for the key (J1) and drain milliammeter (J2). This allows the circuit to restore itself automatically when the key or meter is removed.

To prevent parasitic oscillation, all leads must be as short, rigid, and direct as practicable. The U222 must be mounted clear of wiring, chassis, and all other components, since the gate electrode is internally connected to the metal case of this FET. Resistor R1 is 2 watts, and capacitors C1 and C3 are mica. Selector switch S1 must be ceramic insulated.

The TNT circuit may be used as a doubler (3.5 to 7 MHz, and 7 to 14 MHz) if a driving signal is fed into the gate coil and the drain tank is tuned to twice the driving frequency.

PHONE MONITOR

A tuned diode detector with headphones is often used to monitor radiophone transmissions. The semiconductor diode has the advantage in this application that, at least for large signals, its response is linear, so that it gives a faithful check of the monitored signal. But its output is relatively low, so that the headphone signal cannot compete with room noise.

The solution is to use a headphone amplifier following the detector. Fig. 5-13 shows an arrangement of this sort. The advantage

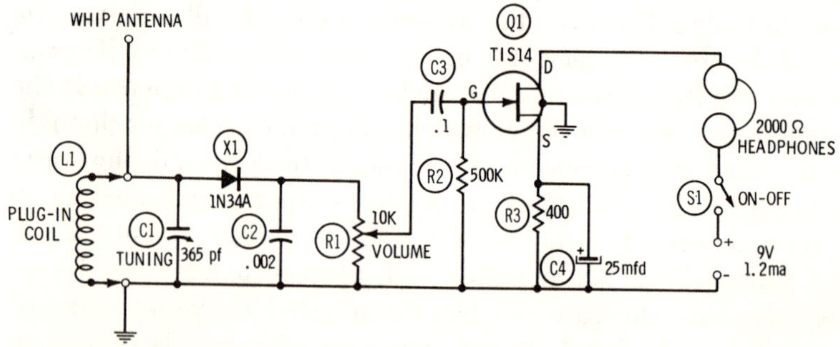

Fig. 5-13. Phone monitor.

of a field-effect transistor (here a TIS14 unit, Q1) is the resultant zero loading of the detector by the amplifier.

A short whip antenna will pick up adequate signal from the transmitter. The signal is tuned in by means of 365-pf variable capacitor C1 which resonates with plug-in coil L1 at the signal frequency. Table 5-2 gives data for coils to cover the frequency range of 0.26 to 100 MHz in five bands. The specified units are

Table 5-2. Phone Monitor Coil Data

Coil No.	Inductance (μh)	Frequency Range (MHz)	Coil Type*
A	1000	0.26—1	40A000CBI
B	110	0.8—3	40A104CBI
C	16	2—8	40A155CBI
D	1.3	7–28	40A156CBI
E	0.1	16—100	40A107CBI

*Catalog numbers of J. W. Miller Co.

miniature, ceramic, slug-adjusted coils which may be set exactly to frequency during calibration of the monitor. For plug-in purposes, these coils may be mounted inside 1 inch-diameter 4-prong forms, such as Millen Type 45004. For narrower-band tuning than is afforded by this coil assortment (example: amateur or citizens' bands), a lower capacitance (for example, 50 or 100 pf) should be used for C1.

The detector is an inexpensive 1N34A germanium diode (X1). Audio output of the diode, developed across load resistor R1, is applied to the gate of the TIS14. Capacitor C3 blocks the d-c component in the output of the diode. Since the load resistor (R1) is a potentiometer, efficient control of headphone volume is assured. Current drain is 1.2 ma at 9 vdc.

Tuning capacitor C1 is a midget, broadcast-type variable. C2 is a mica capacitor, and C3 a 100-volt plastic unit. C4 is a 25-dcwv electrolytic. Resistors R2 and R3 are 1/2 watt.

All wiring must be as short, direct, and rigid as practicable. A fourth pigtail of the TIS14 is internally connected to the metal case of this FET and must be grounded as shown, for shielding.

The monitor may be calibrated by loosely coupling coil L1 to an amplitude-modulated r-f signal generator, tuning in generator signals at selected frequencies by adjusting C1 for loudest headphone response, and graduating the C1 dial accordingly.

6

Control Circuits

The sensitivity of the field-effect transistor to low signal levels and to small changes in level, together with its negligible loading of timing circuits, signal circuits, and actuating devices, suits it to a variety of electronic control circuitry where special provision often must be made for the bipolar transistor. Usually, a simple FET circuit in a control system offers straightforwardness and economy not so easily obtained with either tubes or bipolar transistors.

In solid-state control circuits, the FET allows some components, such as high-resistance photocells and capacitive transducers, to be operated with greater efficiency and convenience than is possible in such circuits containing bipolar transistors. And a special member of the field-effect family—the *photo-FET*, a light-sensitive FET—weds photocell and field-effect transistor.

This chapter describes several basic control circuits which are intrinsically useful. Methods of operating relays from a variety of signals—direct current, radio frequencies, audio frequencies, and illumination levels—are shown. Perhaps these can also serve to generate ideas for other uses.

D-C RELAY

Fig. 6-1 shows the circuit of an amplifier-type d-c relay based on a single 2N3819 FET. This circuit has an input resistance of 10 megohms due entirely to resistor R1, so it draws only 0.15 microampere from the d-c control-signal source.

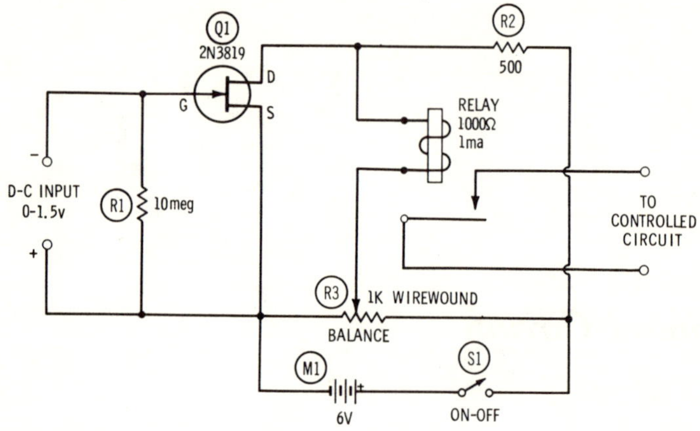

Fig. 6-1. D-c relay.

The 1000-ohm, 1-ma d-c relay (Sigma 5F, or equivalent) is connected in a resistance-bridge circuit whose four arms are R2, the internal drain-to-source resistance of the FET, and the two halves of potentiometer R3. With the D-C INPUT terminals open, R3 is set to balance the static FET current out of the relay (at balance, the relay opens firmly), and the bridge should remain balanced indefinitely without further attention.

Subsequent application of a d-c input signal of 1.5 volts will unbalance the bridge by changing the FET drain-to-source resistance; this will cause approximately 1.2 ma to flow through the relay, closing the latter. Up to 20 megohms may be used at R1, thereby decreasing the drain on the signal source to as little as 0.075 microampere.

All fixed resistors are 1/2 watt. Maximum current drain from the 6-volt d-c supply (M1) is 10 ma. The 2N3819 is a plastic-encapsulated FET which, because of this construction, demands none of the installation precautions occasioned by metal-cased units. The plastic housing will withstand soldering temperatures.

A-C R-F RELAY

The d-c relay described in the preceding section may be actuated by an a-c or r-f control signal if a diode rectifier is operated ahead of the circuit, as shown in Fig. 6-2.

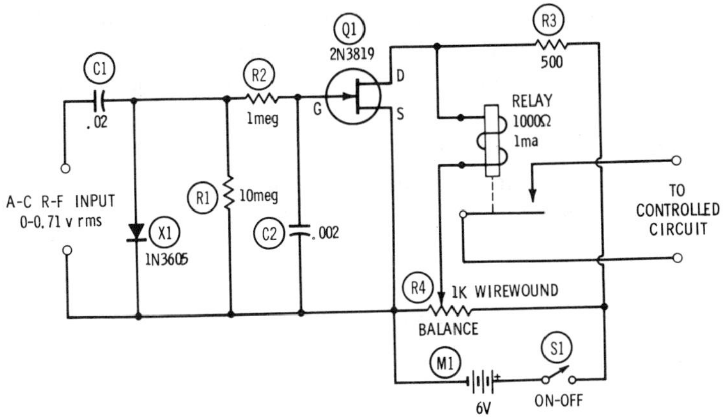

Fig. 6-2. A-c r-f relay.

A shunt-diode rectifier circuit is used (C1-R1-X1). The main advantage of the shunt diode is isolation from any d-c component present in the signal source. In this arrangement, capacitor C1 is charged to the peak value of the a-c signal (less the small forward voltage drop in the diode) on positive half-cycles of signal voltage, and this peak d-c voltage (across 10-megohm load resistor R1) is presented to the gate of the FET. The diode must be polarized as shown for the d-c voltage at the gate to be negative. A silicon diode (1N3605) is required to work into the high load resistance, R1; the back resistance of germanium diodes is too low to support rectification with such a high value of load. Resistor R2 and capacitor C2 form a filter to remove any signal-voltage fluctuations from the diode d-c output.

The 1000-ohm, 1-ma d-c relay (Sigma 5F, or equivalent) is connected in a resistance-bridge circuit whose four arms are R3, the internal drain-to-source resistance of the FET (2N3819), and the two halves of potentiometer R4. With the AC-INPUT terminals open, R4 is set to balance the static FET current out of the relay (at the balance point, the relay opens firmly), and the bridge should then remain balanced indefinitely without further attention.

Subsequent application of an a-c or r-f input signal of 0.71 volt rms will unbalance the bridge by changing the FET drain-to-source resistance, and this will cause approximately 1.2 ma to flow through the relay, closing the latter.

All fixed resistors are 1/2 watt, and capacitors C1 and C2 are mica. Maximum current drain from the 6-volt d-c supply (M1) is 10 ma. The 2N3819 is a plastic-encapsulated FET which, because of this construction, demands none of the installation precautions occasioned by the metal-cased units.

The terminal arrangement of the FET's can be checked against the drawings shown in Appendix A. Manufacturers may be obtained from Appendix B, and the addresses are shown in Appendix C.

TOUCH-PLATE RELAY

The touch-plate relay has many commercial, industrial, and household applications. Touch the metal plate lightly and the relay closes; remove your finger and the relay opens. Call signals, safety devices, intrusion alarms, and kindred devices are operated from such relays.

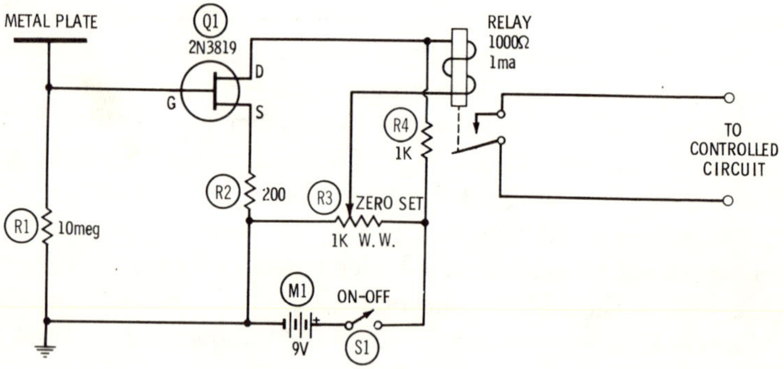

Fig. 6-3. Touch-plate relay.

Fig. 6-3 shows the circuit of a simple touch-plate relay based on one 2N3819 FET. In this arrangement, the touch plate is a disc or plate of sheet metal or foil (2-inch minimum diameter) or a metal ball. Zero Set potentiometer R3 is adjusted until the relay just opens. When the plate is undisturbed, the FET draws negligible current through the relay (Sigma 5F, 1000-ohm, 1-ma, or equivalent). When the plate is touched, enough stray noise is coupled into the 10-megohm gate circuit to boost the drain current to approximately 1.7 ma and close the relay. (This assumes a firm touch; a lighter touch will produce a lower current change, but the relay will close as long as this current reaches a minimum value of 1 ma.

If desired, a delayed dropout may be obtained by means of a capacitor in parallel with the relay coil. With the 1000-ohm coil, the hold-in is approximately 0.8 second per 1000 mfd of parallel capacitance.

Resistors R1 and R2 are 1/2 watt. The 2N3819 is a plastic-encapsulated FET which demands none of the installation precautions occasioned by metal-cased units.

AMPLIFIED PHOTOCELL

The silicon solar cell (self-generating type) is basically a low-voltage, high-current device. In some applications, especially where high impedances are involved, more convenient control might be effected if the d-c output voltage of the cell could be boosted and the cell looked like a high-impedance source.

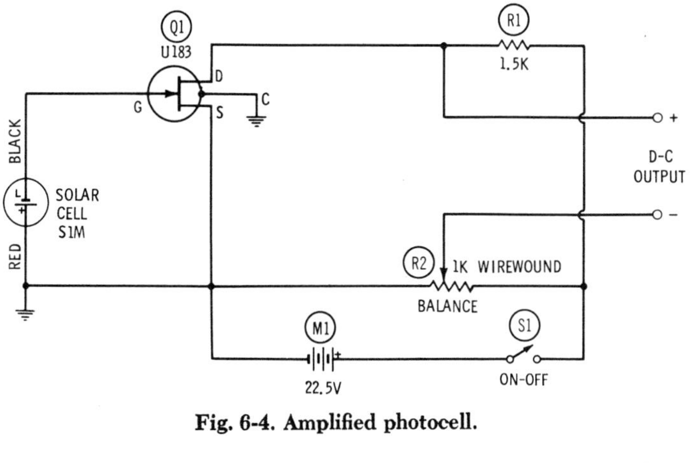

Fig. 6-4. Amplified photocell.

Fig. 6-4 shows the circuit of a simple d-c amplifier based on a single U183 FET. The maximum no-load output (0.4 volt) of the solar cell (International Rectifier S1M, or equivalent) is amplified to 1.6 volts by this arrangement. A particular advantage of a circuit such as this is the high-resistance load which the FET input offers to the cell. On rapidly changing light signals and those modulated at high frequencies, this input avoids the high damping caused by the low resistances into which such silicon cells usually operate.

The output half of the circuit is a resistance bridge whose four arms are R1, the internal drain-to-source resistance of the FET, and the two halves of potentiometer R2. With the solar cell completely darkened, R2 is set to balance the static voltage from the D-C OUTPUT terminals (a d-c vtvm temporarily connected to the output terminals will serve as a balance indicator). At the balance point, the indicator will read zero, and the bridge should then remain balanced indefinitely without further attention.

Subsequent illumination of the cell will unbalance the bridge because the d-c output voltage of the cell changes the drain-to-source resistance of the FET. This will cause 1.6 volts to appear at the D-C OUTPUT terminals. (This level of output voltage will maintain itself across a 20,000-ohm load.)

Wiring is uncritical. Resistor R1 is 1/2 watt, and potentiometer R2 is 1 watt. The fourth pigtail of the U183 is internally connected to the metal case of this FET and should be grounded as shown. Total current drawn from the 22.5-volt source (M1) is 24 ma.

PHOTO-FET RELAY

The photo-FET (photoelectric field-effect transistor) has several advantages over photodiodes and bipolar phototransistors. It combines, for example, the functions of self-generating photojunction and high-impedance–input amplifier. And it provides high amplification at low noise level (0.5-db typical noise).

Fig. 6-5 shows the circuit of a light-sensitive relay employing a P-102 photo-FET. Note that this is a P-type field-effect transistor. The P-102 has a built-in lens, which greatly simplifies application.

In this arrangement, the dark current in the photo-FET drain circuit is approximately 300 μa. When the FET is illuminated at only 10 foot-candles, the drain current increases to approximately 1.6 ma, more than enough to close the 2500-ohm, 1.4-ma d-c relay (Sigma 5F, or equivalent). If desired, a delayed dropout may be obtained by means of a capacitor in parallel with the relay coil.

With the 2500-ohm coil shown, the hold-in is approximately 2.75 seconds per 1000 mfd of parallel capacitance. Current drain on the 1.5-volt gate-bias source, M1, is less than 0.1 microampere.

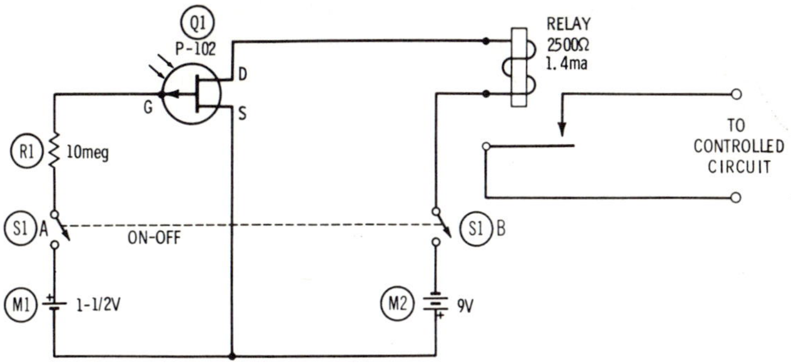

Fig. 6-5. Photo-FET relay.

Wiring is straightforward, and resistor R1 is 1/2 watt. The gate electrode of the P-102 is internally connected to the metal case of this FET, so the latter must be kept clear of contact with wiring, metal chassis, or other components.

TIMER

Resistance-capacitance–controlled timers have been built with vacuum tubes, thyratron tubes, bipolar transistors of all types, and with no tubes or other active devices whatever. The bipolar-transistor circuits, while offering the advantage of compactness, small size, light weight, freedom from the power line, and zero heat generation, have been troublesome to design because the low-impedance transistors load the r-c timing circuit and impose undesirable limitations on it. The high input impedance of the field-effect transistor, however, allows vacuum-tube performance to be attained in solid-state electronic timers.

Fig. 6-6 shows the circuit of a timer employing a single U183 FET. This is essentially a bridge-balanced d-c amplifier, with a relay in its output circuit, which receives its driving signal from a charged capacitor discharging through an adjustable resistor. It provides a hold-in of the relay that ranges from 1 second to 80 seconds, as selected by adjustment of the resistance.

The timing interval is determined by the number of seconds taken by 100-mfd capacitor C1, after being fully charged, to discharge through rheostat R1 to 0.4 of its charged voltage. This capacitor is charged by momentarily depressing push-button switch S1. This connects C1 to a 1.5-volt tap on resistor R2. When released, the push-button blade returns to its normal position against the upper contact, and this connects the charged capacitor to rheostat R1 and the gate of the FET.

The 1000-ohm, 1-ma d-c relay (Sigma 5F, or equivalent) is connected in a resistance bridge whose four arms are R3, the internal drain-to-source resistance of the FET, and the two halves of potentiometer R4. With capacitor C1 completely discharged, R4 is

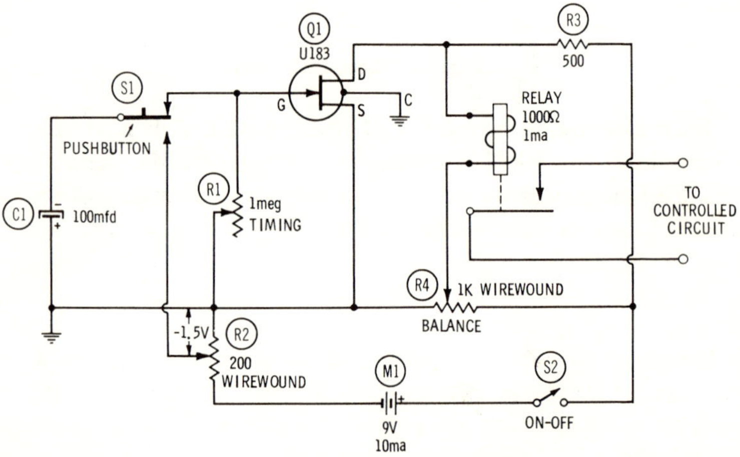

Fig. 6-6. Timer.

set to balance the static FET current out of the relay (at balance, the relay opens firmly), and the bridge then should remain balanced indefinitely without further attention.

To use the timer, depress S1, charging capacitor C1. Then release this switch, whereupon the relay will pick up immediately, since the negative voltage presented to the FET gate by the charged capacitor unbalances the bridge, causing current through the relay. The relay will hold in until the capacitor has discharged sufficiently (through rheostat R1) to reduce the gate voltage to 0.4 of its value at full charge of the capacitor. This time interval depends on the resistance setting of R1, and the rheostat accordingly may be calibrated to read directly in seconds. This may be done with a stopwatch, or the time (t in seconds) may be calculated approximately from the resistance setting (R1 in ohms), the capacitance (C1 in farads), and a multiplier (0.8 for the relay specified here). Thus:

$$t = 0.0001(0.8)R_1 = 0.00008R_1 \text{ (sec)}$$

With a 1-megohm volume control having an audio taper for R1, time intervals from 1 to 80 seconds may be read on a single dial scale.

Various combinations of capacitance and variable resistance other than the 100 mfd and 1 megohm indicated in Fig. 6-6 may be used. In any instance, the electrolytic capacitor used for the purpose should be new and have a good power factor. For close reading or for coverage of narrow bands, several switched-in rheostats may be used against a single capacitor.

Resistor R3 is 1/2 watt; R2 and R4 may be wirewound rheostats with slotted shafts for screwdriver adjustment. Alternatively, R2 may be a stick-type resistor with slider. Electrolytic capacitor C1 is rated at 25 dcwv. The fourth pigtail of the U183 is internally connected to the metal case of this FET and should be grounded as shown.

FET AS VOLTAGE-VARIABLE RESISTOR

Since at any value of drain-to-source d-c voltage, the drain current is a function of the applied gate-to-source bias voltage, the instantaneous value of internal drain-to-source resistance depends on the d-c gate bias. This provides a convenient means of varying resistance with an available variable voltage; and since the correctly polarized gate bias in the FET results in virtually zero gate current, the process is essentially a voltage-controlled one.

Fig. 6-7A shows the simple setup of a U222 (high-current) FET as a voltage-controlled resistor. Here, the control voltage (as gate bias) is applied to terminals A and B. Looking into terminals C and D, one sees a resistor varied by the voltage on A and B. In use, this "resistor" is connected in series with a d-c supply and load device, the supply being polarized so that C is positive and D negative. Some devices in which a voltage-variable resistor is advantageous are voltage regulators, signal translators, voltage-tuned frequency-sensitive r-c networks, voltage-tuned audio oscillators, remote-control devices, temperature transducers, and remote volume controls.

Fig. 6-7B shows performance of the resistor when the voltage between C and D is 15 volts. Note that the resistance varies from

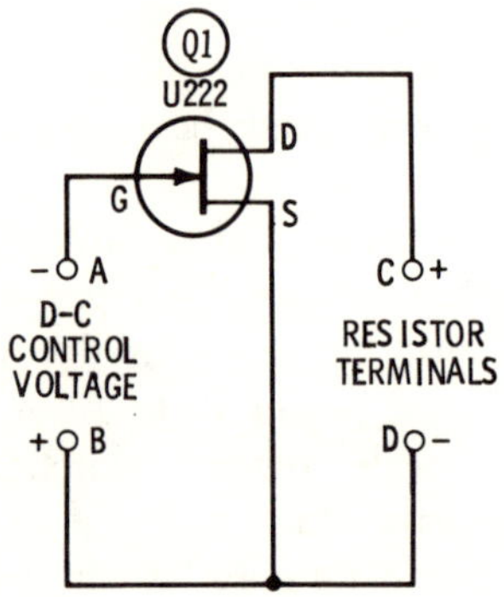

(A) Circuit.

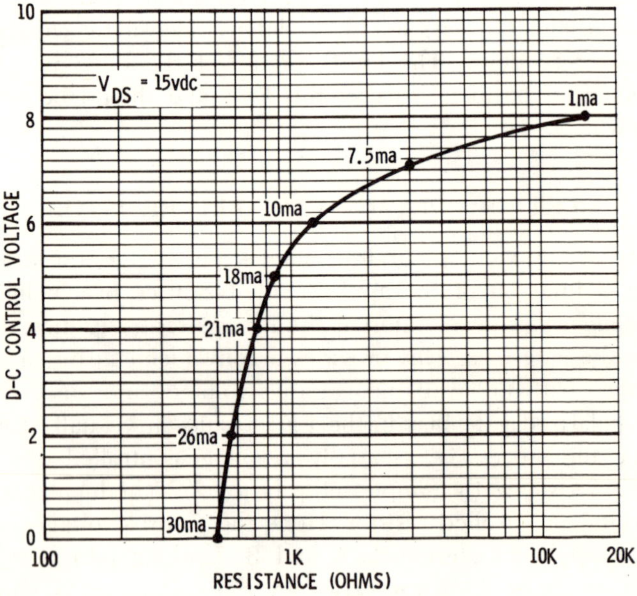

(B) Performance.

Fig. 6-7. FET as voltage-variable resistor.

500 ohms (when the control voltage is zero) to 15,000 ohms (when the control voltage is −8 volts). Other resistance ranges are obtained when the C-D voltage is some value other than 15 volts. The power-handling capability of the variable FET resistor corresponds to the rated power dissipation of the FET. (The manufacturer gives 300 mw as the absolute maximum device dissipation for the U222 at 25°C free-air temperature). Note that the corresponding drain-current values are given along the curve in Fig. 6-7B.

When using any FET as a voltage-variable resistor, keep well within the maximum safe drain-to-source voltage, drain-current, and device-dissipation ratings supplied by the FET manufacturer. The "resistor" may be calibrated by successively applying the various intended values of bias voltage and noting the corresponding drain current. The resistance is calculated at each point as V_D/I_D, where V_D is the instantaneous drain-to-source voltage (in volts) and I_D is the corresponding drain current (in amperes).

CONSTANT-CURRENT SOURCE

The pentodelike drain-voltage/drain-current characteristic curves of the FET suit this component for use as a constant-current adapter. Fig. 6-8 shows the circuit of such an adapter employing a TIX33 FET.

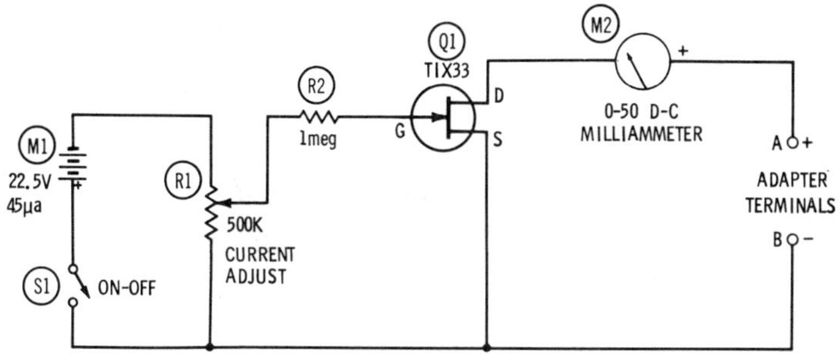

Fig. 6-8. Constant-current source.

The adapter is connected in series with an external d-c power supply and the device operated from the supply, with the supply polarized correctly to make terminal A of the adapter positive and B negative. Potentiometer R1 then is set for the desired load current, as indicated by 0-50 d-c milliammeter M2. (With the TIX33, this may be any value between 1 and 40 ma.) That current level then will be maintained automatically within 6 percent, even if the voltage of the external supply varies between 3 and 20 volts.

Current drain from the 22.5-volt gate bias supply, M1, is 45 microamperes, which is due entirely to current through the 0.5-megohm potentiometer, R1. Resistor R2 is 1/2 watt.

SENSITIVE A-C A-F RELAY

Using two FET's (Q1, Q2) and one silicon bipolar transistor (Q3), the circuit given in Fig. 6-9 closes a 1-milliampere, 5-volt d-c relay (K1) in response to an input signal of only 0.1 volt rms

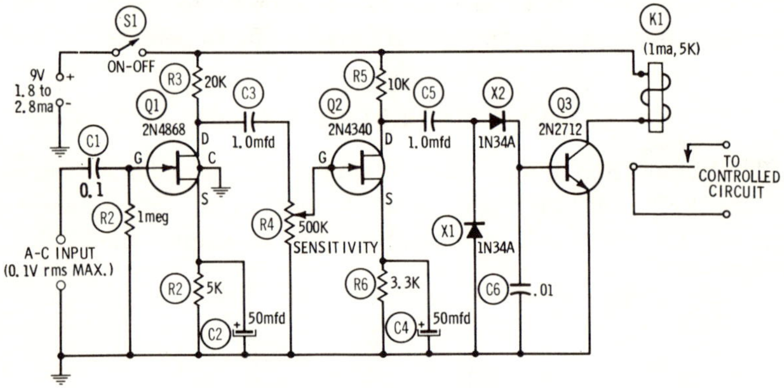

Fig. 6-9. Sensitive a-c a-f relay.

(at approximately 0.1 microampere), when gain control R4 is set to maximum. Good response is obtained throughout the spectrum from 50 Hz to 50 kHz.

The input signal is amplified by a two-stage r-c–coupled section based on two FET's: 2N4868 (Q1) and 2N4340 (Q2). The amplified signal then is rectified by the 1N34A diodes (X1, X2), and the resultant direct current is amplified by a 2N2712 plastic-encapsulated silicon bipolar transistor (Q3) and applied to the 1-ma, 5000-ohm d-c relay (K1, Sigma 5F, or equivalent).

With zero input signal, the 2N2712 draws virtually no current, and the total circuit drain is 1.8 ma at 9 vdc. With a 0.1-volt input signal (R4 set for maximum gain), the circuit drain rises to 2.8 ma.

All fixed resistors are 1/2 watt. Capacitors C1, C3, C5, and C6 are 100-volt plastic units, and C2 and C4 are 25-dcwv electrolytics.

For stability, all wiring must be kept as short, direct, and rigid as practicable, and the entire unit should be enclosed in a grounded shield box if interfering fields are present. A fourth pigtail of the 2N4868 is internally connected to the metal shell of this FET and must be grounded as shown, for shielding. In the 2N4340, the gate electrode is internally connected to the metal case, so this FET must be mounted clear of chassis, wiring, and other components. Signal input leads to the A-C INPUT terminals should be shielded.

SOUND-OPERATED RELAY

Fig. 6-10 shows the circuit of a sound-operated relay based on two 2N4868 FET's (Q1, Q2) and one U183 FET (Q3). With this arrangement, a microphone output signal of 1.8 millivolts rms (−52.5 db) will close the 1-ma, 1000-ohm d-c relay (K1) when sensitivity control R4 is set to maximum.

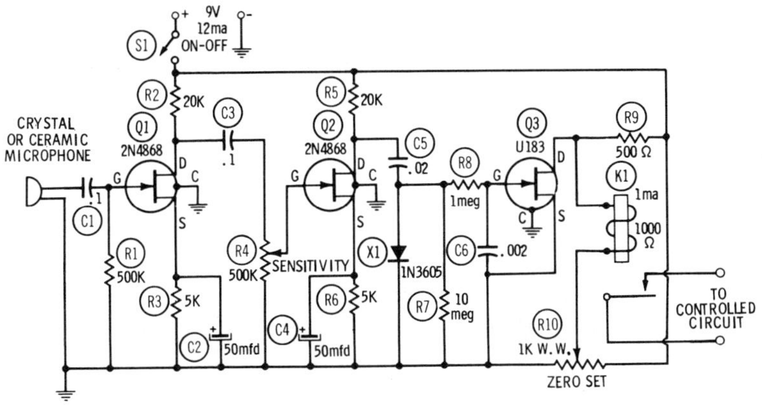

Fig. 6-10. Sound-operated relay.

The first two stages form a high-gain r-c–coupled amplifier, the output of which is rectified by the 1N3605 silicon diode, X1. The resultant d-c output voltage of the diode, developed across load resistor R7, is applied to the gate of the U183 FET (Q3) through the resistance-capacitance filter R8-C6. The U183 acts as a d-c amplifier which drives a 1-ma, 1000-ohm d-c relay (Sigma 5F, or equivalent). Current drain is 12 ma at 9 vdc.

Initial adjustment of the circuit consists of balancing the static drain current of the U183 out of the relay:

1. Short-circuit the microphone terminals.
2. Set potentiometer R4 for maximum sensitivity.
3. Adjust potentiometer R10 until relay K1 opens.

(This adjustment may be made very precisely if a d-c vtvm is connected across the relay coil and R10 is adjusted for zero deflection of the meter.)

All fixed resistors are 1/2 watt. Potentiometer R10 is wirewound. Capacitors C1, C3, C5, and C6 are 100-volt plastic units, and C2 and C4 are 25-dcwv electrolytics. The high-impedance input of the first stage will accommodate a crystal or ceramic microphone directly; for a dynamic microphone, use an appropriate input transformer.

For stability and minimum stray signal pickup, all wiring must be as short, direct, and rigid as practicable. For complete shielding in electrically noisy locations, the entire unit must be enclosed in a grounded metal box. In each of the FET's, a fourth pigtail is internally connected to the metal case and must be grounded as shown, for shielding.

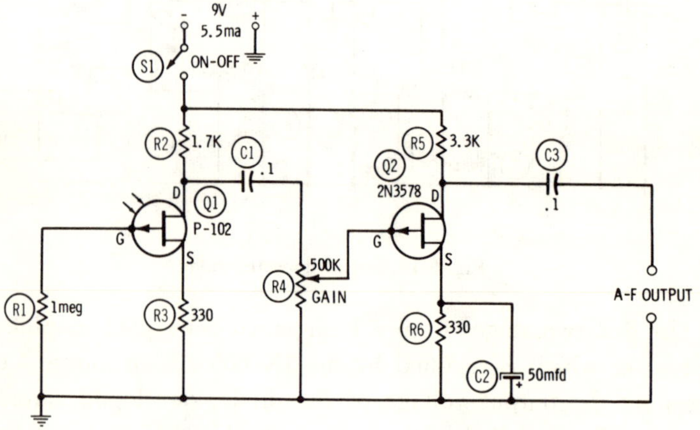

Fig. 6-11. Modulated light detector/amplifier.

MODULATED-LIGHT DETECTOR/AMPLIFIER

The P-102 photo-FET is a good detector of modulated light (either chopped or smoothly modulated), and its a-f output is easily amplified. Fig. 6-11 shows a circuit composed of a P-102 photo-FET (Q1) and a 2N3578 P-channel FET (Q2) as an a-f amplifier.

The sensitivity of this circuit is such that 15 foot-candles of light modulated or chopped at a 1000-Hz rate will produce 1 volt rms (no load) at the A-F OUTPUT terminals when potentiometer R4 is set for maximum gain. Current drain is 5.5 ma at 9 vdc.

The detector/amplifier has many uses, such as light-beam communication, alarm signaling, intrusion alarm service, sound-on-film pickup, and optoelectronic control. The virtually zero loading of the photo-FET by the amplifier results in maximum efficiency of the P-102, while the amplifier produces a useful output-signal amplitude for direct application or further amplification.

All fixed resistors are 1/2 watt. Capacitors C1 and C3 are 100-volt plastic units, and C2 is a 25-dcwv electrolytic. In both the P-102 and the 2N3578, the gate electrode is internally connected to the metal case of the FET, so each of these units must be mounted clear of other components, chassis, and wiring. The P-102 must be positioned so that its nose-end lens faces the incoming light. All wiring must be as short, rigid, and direct as practicable.

PHASE SHIFTER

The circuit given in Fig. 6-12, combining a triode phase inverter and an adjustable r-c circuit, provides continuously variable phase shift from 0 to 180 degrees while leaving the signal amplitude virtually untouched.

This arrangement, based on one 2N3578 FET (Q1), has equal fixed resistances (R2 and R3, matched within 1 percent) in both drain and source legs. When rheostat R4 is set to one end of its range, the output is taken from the source of Q1 and is in the same phase as the input signal (Q1 is then acting as a source follower); therefore, there is zero phase shift between output and input. But when R4 is set to the opposite end of its range, the output is taken from the drain of Q1 and is of opposite phase from the input (Q1 then is acting as a conventional triode amplifier); therefore, there is then 180 degrees phase shift between output and input. At intermediate settings of R4, the various phase angles between 0 and 180 degrees are obtained. Two identical stages may be cascaded for 0 to 360 degrees phase shift.

The circuit provides good, full-range operation between 150 Hz and 5 kHz. Below 150 Hz and above 5 kHz, inherent phase shifts in the circuit and components restrict the range to limits somewhat higher than zero degrees on one end and somewhat lower than 180 degrees on the other. However, some improvement may be obtained by careful attention to lead dress and shortness of leads, and to minimizing stray capacitances.

For undistorted operation, the maximum input-signal amplitude is 2 volts rms. With potentiometer R1 set for maximum gain the corresponding output-signal amplitude is 0.75 volt rms. Current drain is 1 ma at 9 vdc.

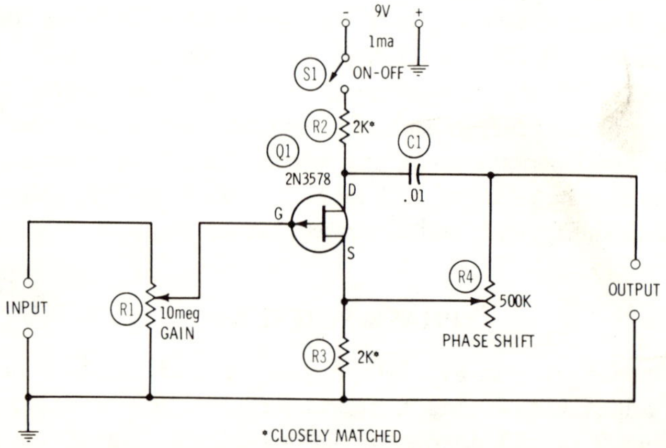

Fig. 6-12. Phase shifter.

Resistors R2 and R3 are 1 watt and are of the 1 percent noninductive, wirewound type. Capacitor C1 is silvered mica.

All wiring must be as short, rigid, and direct as practicable and kept well away from the metal chassis if one is used. The input and output circuits must be kept well separated. In the 2N3578, the gate electrode is internally connected to the metal case, so this FET must be mounted clear of other components, wiring, and chassis.

The dial of the phase-shift control, R4, may be calibrated in degrees by means of Lissajous figures. For this purpose, connect an audio-frequency signal generator to the INPUT terminals and to the vertical-amplifier input terminals of an oscilloscope having excellent phase characteristics, and connect the OUTPUT terminals of the phase shifter to the horizontal-amplifier input of the oscilloscope.

7

Instrument Circuits

Solid-state test equipment has long been prized for its complete portability, electrically clean operation, and total isolation from the power line. In all classes of such equipment, the bipolar transistor, often in company with the semiconductor diode, brought about many improvements. However, its use generally necessitated design compromises. But the FET, being more nearly like the vacuum tube, demands few compromises in the transistorizing of standard instrument circuits.

The electronic voltmeter is a clear example of successful transistorization via the field-effect transistor. Here, the FET has faithfully preserved the high input resistance (1 megohm or more per volt) which is characteristic of the vacuum-tube voltmeter, whereas transistorization via the bipolar transistor usually afforded an input resistance of only 100,000 ohms per volt. Another example is the highly useful dip oscillator; in this instrument, the FET provides the same high sensitivity and excellent selectivity that is characteristic of the grid-dip oscillator.

This chapter describes several simple instruments which are made efficient, reliable, and completely portable by the FET. For circuits in other chapters, which may be used for instrumentation, see Figs. 2-11, 2-12, 2-13, 3-1, 3-2, 3-3, 3-4, 3-7, 3-8, 3-9, 3-10, 4-7, 5-10, and 6-6.

SINGLE-FET ELECTRONIC D-C VOLTMETER

Fig. 7-1 shows a transistorized version of the extremely useful d-c vacuum-tube voltmeter. This instrument, which (like its vtvm counterpart) has a constant input resistance of 11 megohms on all ranges, employs a single 2N4340 FET. (1 megohm of the input resistance is accounted for by isolating resistor R1 in the probe.)

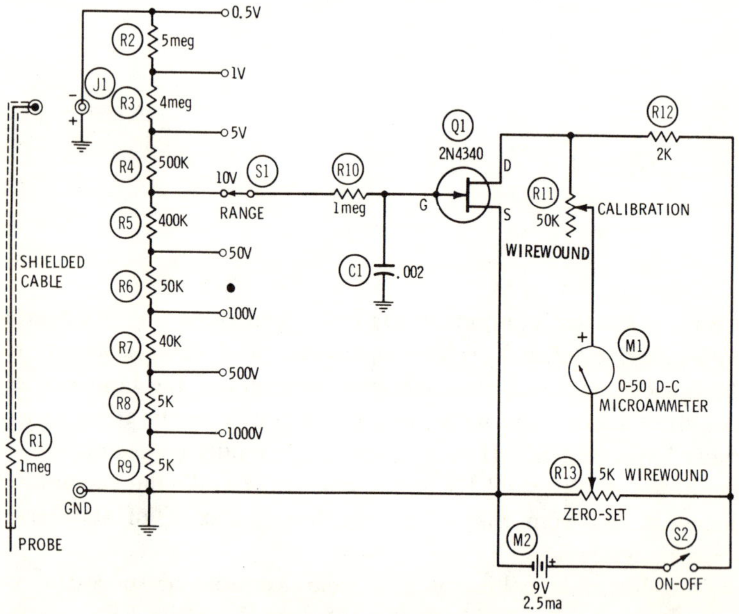

Fig. 7-1. Electronic d-c voltmeter (single FET).

The instrument covers 0 to 1000 volts in eight ranges: 0-0.5, 0-1, 0-5, 0-10, 0-50, 0-100, 0-500, and 0-1000 selected with single-pole, 8-position, nonshorting rotary switch S1. The range resistors (R2 to R9) in the input voltage-divider string must be accurate to at least 1 percent. Some of these (400K and 4 meg) are not stock values and must be made up by series-connecting available resistors (300K + 100K, and 3 meg + 1 meg, respectively). Linearity of response is within ±2 percent of full scale.

Resistor R10 and capacitor C1 form a filter to remove any stray fluctuations picked up by the test leads. Rheostat R11 is the CALIBRATION control mounted safely inside the instrument case, and potentiometer R13 is the front-panel ZERO-SET.

The 0-50–microampere scale of meter M1 reads direct on the 50-volt range and is read on the 0.5-, 5-, 50-, and 500-volt ranges by mentally shifting the decimal point, as required. On the 1-, 10-, 100-, and 1000-volt ranges, however, the readings must be mentally multiplied by 2 or some multiple of 2 (by 0.02 for the 1-volt, by 0.2 for the 10-volt, by 2 for the 100-volt, and by 20 for the 1000-volt range).

To calibrate the instrument:

1. Set switch S1 to the 1-volt range.
2. Set rheostat R11 approximately to the center of its range.
3. Plug the test leads into jack J1 and the GND terminal.
4. Close switch S2.
5. Adjust potentiometer R13 to zero the meter.
6. Connect an accurately known 1-volt d-c source to the input (negative to the shielded test lead, positive to the GND terminal).
7. Adjust rheostat R11 for exact full-scale deflection of meter M1. This rheostat then need not be touched again until the instrument is routinely recalibrated.

A diode rectifier operated ahead of the input voltage divider will adapt the instrument for a-c voltage measurements but will require special meter scales and calibrations, in addition to the d-c ones.

All fixed resistors are 1 watt, and variable resistors R11 and R13 both are wirewound (the former is provided with a slotted shaft for screwdriver adjustment). Capacitor C1 may be mica or ceramic. The 2N4340 must be kept clear of contact with the chassis or other components, since the gate electrode is internally connected to the metal case of this FET. Current drain from the 9-volt battery, M2, is 2.5 ma.

BALANCED ELECTRONIC D-C VOLTMETER

As in vacuum-tube voltmeters, a symmetrical (balanced) FET d-c voltmeter is superior to the single-ended circuit described in the

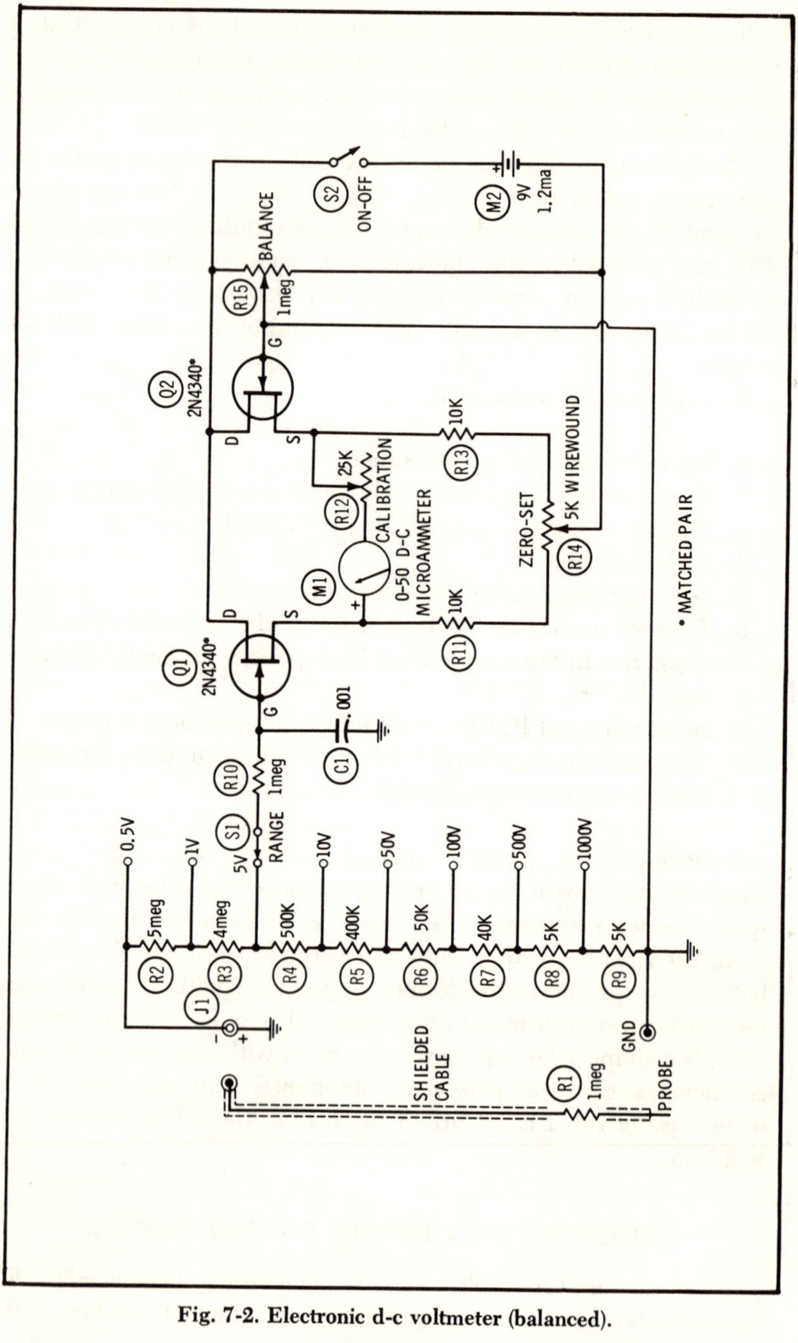

Fig. 7-2. Electronic d-c voltmeter (balanced).

preceding section. The balanced circuit exhibits less temperature drift, and therefore requires fewer zero resettings than in the simpler circuit. It also draws less current from the power supply during normal operation.

Fig. 7-2 shows a balanced d-c voltmeter circuit employing two 2N4340 FET's. For best results, these FET's must be matched, and may be so obtained from the manufacturer. The circuit is essentially that of a resistance bridge, the four arms of which are (1) the internal drain-to-source resistance of FET Q1, (2) the internal drain-to-source resistance of FET Q2, (3) R11 and that part of R14 to the left of the slider, and (4) R13 and that part of R14 to the right of the slider. Adjustment of potentiometer R14 balances the bridge and thus zeroes the meter. Application of a test voltage to the gate of Q1 (through the input terminals of the instrument) changes the internal resistance of Q1, unbalancing the bridge, and causes the meter to deflect.

The instrument covers 0 to 1000 volts in eight ranges: 0-0.5, 0-1, 0-5, 0-10, 0-50, 0-100, 0-500, and 0-1000 volts selected with single-pole, 8-position, nonshorting rotary switch S1. The range resistors (R2 to R9) must be accurate to at least 1 percent. Some of these (400K and 4 meg) are not stock values and must be made up by series-connecting available resistors (300K + 100K, and 3 meg + 1 meg, respectively). Linearity of response is within ±2 percent of full scale. The input resistance of the instrument is constant at 11 megohms on all ranges (1 megohm of this is accounted for by the isolating resistor, R1, in the test probe).

Resistor R10 and capacitor C1 form a filter to remove any stray fluctuations picked up by the test leads. Rheostat R12 is the CALIBRATION control mounted safely inside the instrument case, and potentiometer R14 is the front-panel ZERO-SET. Potentiometer R15 (also mounted safely inside the case) is used to balance the circuit by varying the grid bias on Q2 before the instrument is placed into service.

The 0-50–microampere scale of meter M1 reads direct on the 0-50–volt range and is easily read on the 0.5-, 5-, 50-, and 500-volt ranges by mentally shifting the decimal point, as required. On the 1-, 10-, 100-, and 1000-volt ranges, however, the reading must be multiplied by 2 or some multiple of 2 (by 0.02 for the 1-volt, by 0.2 for the 10-volt, by 2 for the 100-volt, and by 20 for the 1000-volt range).

The circuit must be initially balanced in the following manner:

1. Unplug the test leads from jack J1 and GND.
2. Set switch S1 to its 0.5-volt range.
3. Close switch S2.
4. Adjust potentiometer R14 to zero the meter, changing the setting of rheostat R12 if necessary to get a readable deflection of the meter when unzeroed.
5. Temporarily ground the gate of FET Q1; this should drive the meter from zero. Without touching R14, adjust R15 to rezero the meter.
6. Remove the ground from the Q1 gate; this should drive the meter from zero. Without touching R15, reset R14 to rezero the meter.
7. Again, ground the gate of Q1 and note if the meter reads off zero. If it does, repeat Step 5.
8. Continue to alternate between these steps until the zero setting of the meter is undisturbed by grounding the gate of Q1. Potentiometer R15 then should be protected from tampering.

After this initial balancing adjustment, the instrument may be calibrated in the following manner:

1. Set switch S1 to the 1-volt range.
2. Set rheostat R12 to approximately the center of its range.
3. Plug the test leads into jack J1 and the GND terminal.
4. Close switch S2.
5. Adjust potentiometer R14 to zero the meter.
6. Connect an accurately known 1-volt d-c source to the input (negative to the shielded test lead, positive to the GND terminal).
7. Adjust rheostat R12 for exact full-scale deflection of meter M1. This rheostat then need not be touched again until the meter is routinely recalibrated.

A diode rectifier may be operated ahead of the input voltage divider to adapt the instrument to a-c voltage measurements. However, this will require special meter scales and calibrations, in addition to the d-c ones.

All fixed resistors are 1 watt, variable resistors R12 and R14 both are wirewound, and variable resistor R15 (because of its high resistance) is a composition type. R12 and R15 each should have a slotted shaft for screwdriver adjustment. Capacitor C1 may be mica or ceramic. The 2N4340's must be kept clear of contact with each other or with the chassis, wiring, or other components, since the gate electrode is internally connected to the metal case of this FET. Current drain from the 9-volt battery, M2, is 1.2 ma.

ELECTRONIC A-C VOLTMETER

Fig. 7-3 shows the circuit of a high-impedance electronic a-c voltmeter suitable for audio and ultrasonic tests and measurements. This circuit, which is based on a single 2N4868 FET, has an input impedance of 1 megohm in series with 0.1 microfarad on all voltage ranges.

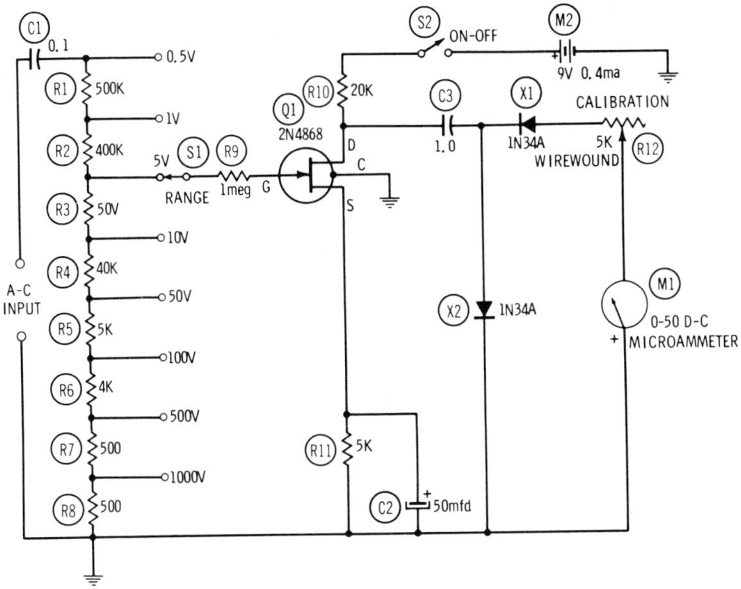

Fig. 7-3. Electronic a-c voltmeter.

The instrument covers 0 to 1000 volts rms in eight ranges: 0-0.5, 0-1, 0-5, 0-10, 0-50, 0-100, 0-500, and 0-1000, selected with single-pole, 8-position, nonshorting rotary switch S1. The range resistors

(R1 to R8) must be accurate to at least 1 percent. Two of these (400K and 4 meg) are not stock values and must be made up by series-connecting available resistors (300K + 100K, and 3 meg + 1 meg, respectively). Linearity of response is within 5 percent of full scale. Frequency response, referred to 1 kHz, is down 3.5 db at 50 Hz and 2 db at 50 kHz.

The circuit is the conventional amplifier-rectifier arrangement: the signal is amplified first by the FET and then is rectified by the two germanium diodes (X1, X2). The d-c output of the diodes is more than sufficient to drive meter M1 to full scale when a signal voltage of 0.5 volt rms is applied to the A-C INPUT terminals. The meter deflection is proportional to the average value of the a-c signal voltage, but the meter may be calibrated to read rms voltage on a sine-wave basis.

No zero adjustment is needed. Rheostat R12, the CALIBRATION control, is mounted safely inside the instrument case to protect it from tampering after the instrument has been calibrated.

The 0-50–microampere scale of the meter reads direct on the 0-50–volt range, and is easily read on the 0.5-, 5-, 50-, and 500-volt ranges by mentally shifting the decimal point, as required. On the 1-, 10-, 100-, and 1000-volt ranges, however, the reading must be mentally multiplied by 2 or some multiple of 2 (by 0.02 for the 1-volt, by 0.2 for the 10-volt, by 2 for the 100-volt, and by 20 for the 1000-volt range).

Calibrate the instrument in the following manner:

1. Set switch S1 to the 5-volt range.
2. Set rheostat R12 to full resistance.
3. Close switch S2.
4. Connect an accurately known 5-volt rms a-c source to the A-C INPUT terminals (any frequency between 60 Hz and 1 kHz).
5. Adjust rheostat R12 for exact full-scale deflection of meter M1. This rheostat then need not be touched again until the instrument is routinely recalibrated.

All fixed resistors are 1 watt. Capacitor C1 is a 1000-volt ceramic unit (Sprague 5HK-P10, or equivalent), C2 a 25-dcwv electrolytic, and C3 a 200-volt metallized paper tubular (Aerovox P8292ZN, or equivalent). Rheostat R12 is a wirewound unit which has a slotted shaft for screwdriver adjustment. The fourth pigtail of the 2N4868 is internally connected to the metal case of this FET and must be grounded as shown. Current drain from the 9-volt battery, M2, is 0.4 ma.

ELECTRONIC A-C VOLTMETER/MILLIVOLTMETER

For more sensitive a-f and ultrasonic voltage measurements than are permitted by the circuit described in the preceding section, the two-transistor circuit shown in Fig. 7-4 is recommended. The additional amplification available in this latter arrangement permits full-scale deflections down to 0.1 volt (100 mv) rms.

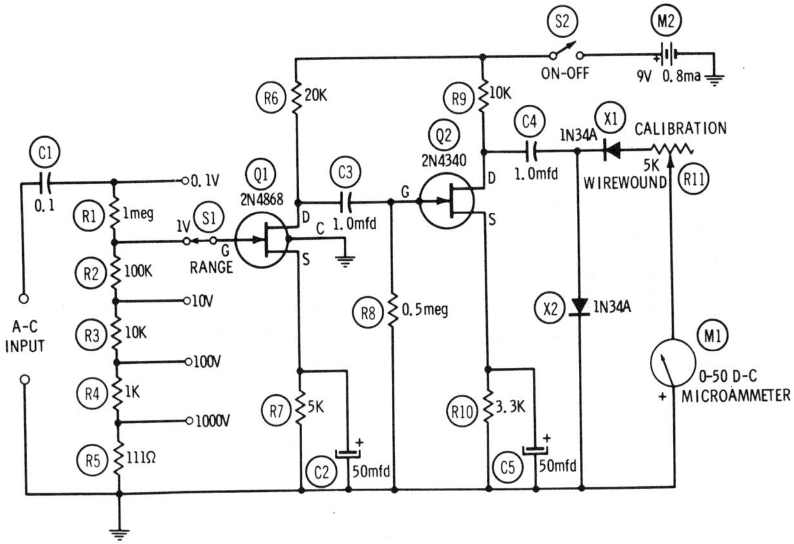

Fig. 7-4. A-f voltmeter/millivoltmeter.

Based on one 2N4868 FET (Q1) and one 2N4340 FET (Q2), the instrument covers 0 to 1000 volts rms in five ranges: 0-0.1, 0-1, 0-10, 0-100, and 0-1000, selected with the single-pole, 5-position, nonshorting rotary switch S1. The range resistors (R1 to R5) must be accurate to at least 1 percent. One of these (R5) is not a stock value and must be made up by series-connecting one 100- and one 11-ohm resistor. (This odd resistance makes it possible to hold the other four resistors to stock values.) Frequency response, referred to 1 kHz, is down 4 db at 50 Hz and 3 db at 50 kHz. Linearity is within 5 percent of full scale on each range.

Like the simpler a-c voltmeter (Fig. 7-3), this instrument uses the conventional amplifier-rectifier arrangement: the signal is amplified by the two FET's (Q1, Q2), then is rectified by two germanium diodes (X1, X2) whose d-c output drives microammeter M1. Deflection of the meter is proportional to the average value of the a-c signal voltage, but the meter may be calibrated to read rms voltage on a sine-wave basis.

No zero adjustment is needed. Rheostat R11, the CALIBRATION control, is mounted safely inside the instrument case to protect it from tampering after the meter has been calibrated.

Unless special scales are drawn for the meter, voltages must be determined from the 0-50–microampere scale by mentally multiplying the deflection by some multiple of 2 (by 0.002 for the 0.1-volt, by 0.02 for the 1 volt, by 0.2 for the 10-volt, by 2 for the 100-volt and by 20 for the 1000-volt range).

Calibrate the instrument in the following manner:

1. Set switch S1 to its 10-volt range.
2. Set rheostat R11 to full resistance.
3. Close switch S2.
4. Connect an accurately known 10-volt rms a-c source to the A-C INPUT terminals (any frequency between 60 Hz and 1 kHz).
5. Adjust rheostat R11 for exact full-scale deflection of meter M1. The rheostat then need not be touched again until the meter is routinely recalibrated.

All fixed resistors are 1 watt. Capacitor C1 is a 1000-volt ceramic unit (Sprague 5HK-P10, or equivalent), C3 is a 200-volt metallized paper tubular, and C2 and C5 are 25-dcwv electrolytics. Rheostat R11 is a wirewound unit which has a slotted shaft for screwdriver adjustment. The fourth pigtail of the 2N4868 (Q1) is internally connected to the metal case of this FET and must be grounded as shown, but the gate electrode of the 2N4340 (Q2) is internally connected to the metal case of the latter FET, which accordingly must be kept clear of contact with wiring, chassis, and other components. Current drain from the 9-volt battery, M2, is 0.8 ma.

The terminal arrangement for the FET's is shown in Appendix A. Manufacturers of FET's can be obtained from Appendix B, and the addresses of the manufacturers are shown in Appendix C.

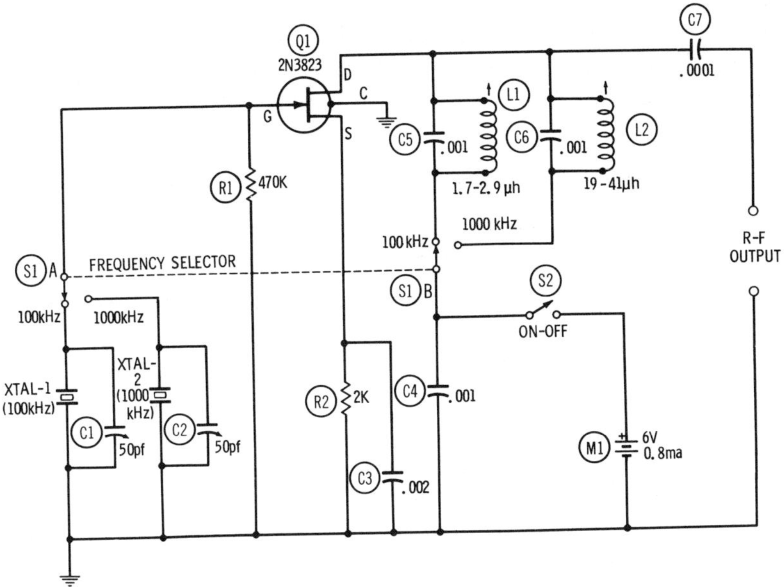

Fig. 7-5. Crystal dual-frequency standard.

CRYSTAL DUAL-FREQUENCY STANDARD

Fig. 7-5 shows the circuit of a dual crystal oscillator with output selectable at either 100 or 1000 kHz. Both signals give strong harmonics for calibration purposes far into the high-frequency spectrum. The circuit is based on a single 2N3823 FET.

A dpdt FREQUENCY SELECTOR switch (S1A-S1B) connects the 100-kHz crystal and the first tank (L1-C5) into the circuit for 100-kHz output, or the 1000-kHz crystal and the second tank (L2-C6) into the circuit for 1000-kHz output. The tank coils are slug tuned for maximum output signal, as indicated by an a-c vtvm or transistorized electronic a-c voltmeter connected to the appropriate output terminals. (L1 is a Miller No. 42A223CBI, or equivalent; L2 is a Miller No. 42A335CBI, or equivalent.) Each crystal frequency is adjustable over a narrow range by means of a 50-pf trimmer capacitor (C1, C2) to standardize the outputs against WWV transmissions or other standard-frequency signals.

All wiring must be kept as short, direct, and rigid as practicable. Also, wiring should be kept well away from metal chassis or case, to minimize stray capacitance which will bypass higher harmonics of the signals. Capacitors C3, C4, and C7 are mica; C5 and C6 are silvered mica. The fourth pigtail of the 2N3823 is internally connected to the metal case of this FET and must be grounded as shown. Current drain from the 6-volt battery, M1, is 0.8 ma.

100-kHz/10-kHz SECONDARY FREQUENCY STANDARD

The secondary frequency standard circuit shown in Fig. 7-6 has a 100-kHz oscillator and 10-kHz multivibrator. This arrangement provides 100-kHz harmonics far into the high-frequency spectrum, and each 100-kHz interval is subdivided by harmonics of the multivibrator. The oscillator is based on a single 2N3823 FET (Q1) and the multivibrator on two 2N4340 FET's (Q2, Q3).

The oscillator is tuned to the crystal frequency by adjusting the slug-tuned coil, L1 (Miller No. 42A223CBI, or equivalent) and the crystal frequency is adjustable over a narrow range, by means of 50-pf trimmer capacitor C1, to standardize the 100 kHz against WWV transmissions or other standard-frequency signals. The 100-kHz output is capacitance coupled through capacitor C3.

Synchronizing 100-kHz voltage is injected into the multivibrator through potentiometer R5 which, with R4, forms the gate resistor of FET Q2. The 10-kHz output is capacitance coupled through capacitor C6.

The multivibrator is best adjusted by connecting the 100-kHz output of the instrument to a communications receiver, tuning in two adjacent 100-kHz harmonics on the receiver, closing switch S1, and adjusting potentiometer R5 until exactly nine intermediate (10-kHz) signals can be counted between each two adjacent 100-kHz harmonics. Adjustment of R5 will cause the intermediate points to jump abruptly from one number to another as the multivibrator synchronizes at 7, 8, 9, 10, or 11 kHz. When the 10-kHz harmonics are not wanted, switch S1 may be opened, and only the 100-kHz harmonics will remain.

The 10-kHz fundamental is separately available, through capacitor C6, as a highly accurate audio frequency, as a square wave, and as a synchronizing signal.

All wiring must be kept as short, direct, and rigid as practicable. Also, wiring should be kept well away from metal chassis or case, to minimize stray capacitance which will bypass higher harmonics of the signals. All fixed resistors are 1/2 watt, and all fixed capacitors

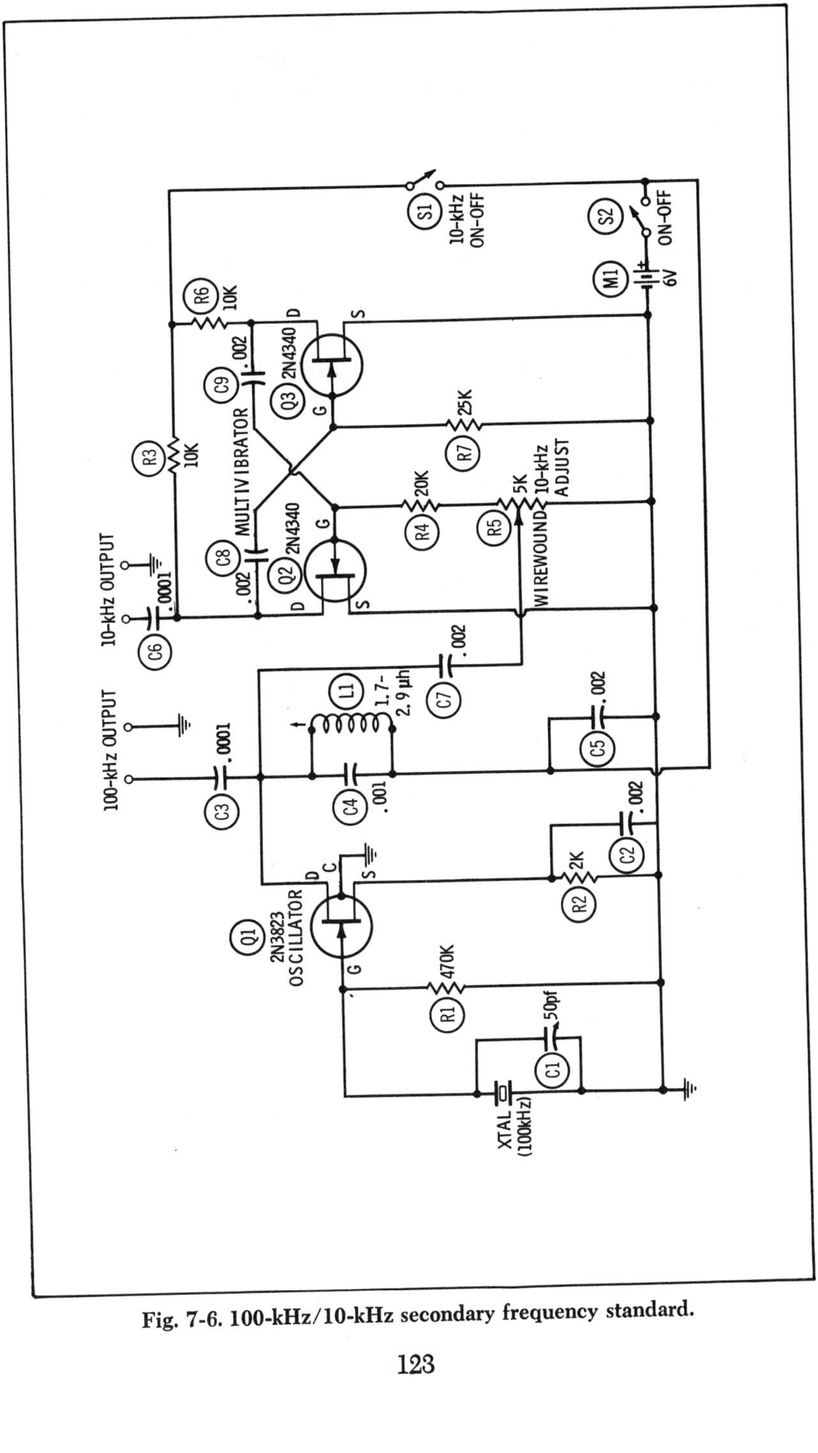

Fig. 7-6. 100-kHz/10-kHz secondary frequency standard.

are silvered mica. Potentiometer R5 is wirewound and provided with a slotted shaft for screwdriver adjustment; it should be mounted safely inside the instrument case to protect it from tampering. The fourth pigtail of the 2N3823 is internally connected to the metal case of this FET and must be grounded as shown. The gate electrode of the 2N4340 is internally connected to the metal case in the latter FET, however, so both 2N4340's must be kept clear of contact with each other or with wiring, chassis, or other components. Total current drain from the 6-volt battery, M1, is 2.2 ma.

DIRECT-READING AUDIO-FREQUENCY METER

A direct-reading audio-frequency meter is a convenient instrument. Feed an unknown audio frequency into its input terminals and read the frequency directly from the meter scale. No adjustments of any kind need be made; only the frequency range must be switched in the same way that a voltmeter range is switched.

Fig. 7-7 shows the circuit of such an instrument, based on one 2N4868 FET (Q1) and one 2N4340 FET (Q2). This instrument covers 0 to 100 kHz in four ranges: 0-100 Hz, 0-1 kHz, 0-10 kHz, and 0-100 kHz. The frequency is indicated on the scale of microam-

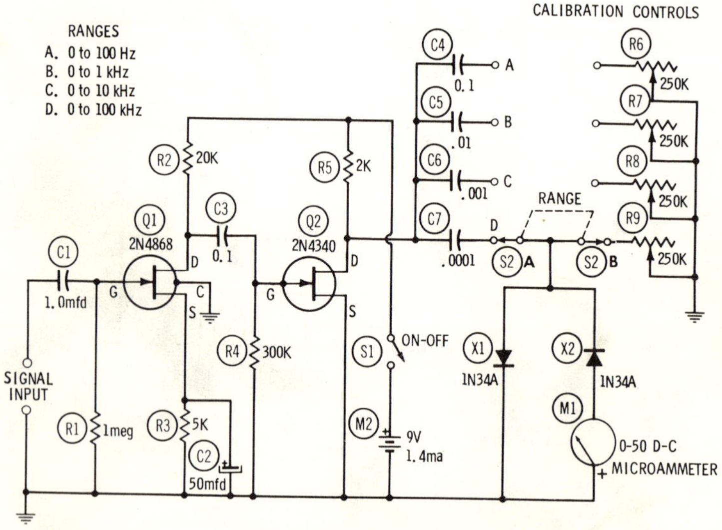

Fig. 7-7. Direct-reading audio-frequency meter.

meter M1, and the indication is independent of signal amplitude from 1.7 volts rms upward and is independent of waveform over a wide range. The response is linear; hence, only one point need be calibrated in each frequency range.

The circuit exploits the same principle employed in some tube-type frequency meters: It is essentially two overdriven amplifier stages in cascade. The output of the last stage accordingly is a square wave which is applied to an r-c circuit (R6-R9 and C4-C7) and diode rectifiers X1 and X2. Since the square wave is limited (i.e., of constant amplitude), the deflection of meter M1 depends only on the *number* of pulses passing through this meter during each second, so is directly proportional to the pulse frequency.

The frequency meter must be calibrated at one point in each of its four ranges (the best point is the top frequency in each range). Rheostats R6, R7, R8, and R9 are the CALIBRATION controls, and these units are provided with slotted shafts for a screwdriver. They are also mounted safely inside the instrument case to protect them from tampering. Follow this calibration procedure:

1. Close switch S1.
2. Set range switch S2 to position A.
3. Connect an accurate audio oscillator to the SIGNAL INPUT terminals and set its output to 2 to 3 volts rms.
4. Set the signal frequency to 100 Hz and adjust R6 for full-scale deflection of meter M1.
5. Set S2 to position B.
6. Set the signal frequency to 1 kHz and adjust R7 for full-scale deflection of the meter.
7. Set S2 to position C.
8. Set the signal frequency to 10 kHz and adjust R8 for full-scale deflection of the meter.
9. Set S2 to position D.
10. Set the signal frequency to 100 kHz and adjust R9 for full-scale deflection of the meter.

All wiring must be kept as short, direct, and rigid as practicable. All fixed resistors are 1/2 watt. Capacitors C1 and C3 are 400-volt metallized paper tubulars, and C2 is a 25-dcwv electrolytic. Capacitors C4 and C5 are 200-volt metallized paper tubulars, and C6 and C7 are silvered mica. While the range capacitors (C4 to C7) should be reasonably accurate, any deviation in capacitance will be compensated by adjustment of the calibration controls (R6 to R9). The fourth pigtail of the 2N4868 is internally connected to the metal case of this FET and must be grounded as shown. In the 2N4340, however, the metal case is internally connected to the gate electrode, so this latter FET must be kept clear of contact with wiring, other components, or chassis. Current drain from the 9-volt battery, M2, is 1.4 ma.

4.5-MHz SOUND-MARKER GENERATOR

Fig. 7-8 shows the circuit of a 4.5-MHz crystal oscillator for sound-channel alignment in tv receivers. This oscillator, based on a single 2N3823 FET, is tuned to the crystal frequency by adjusting

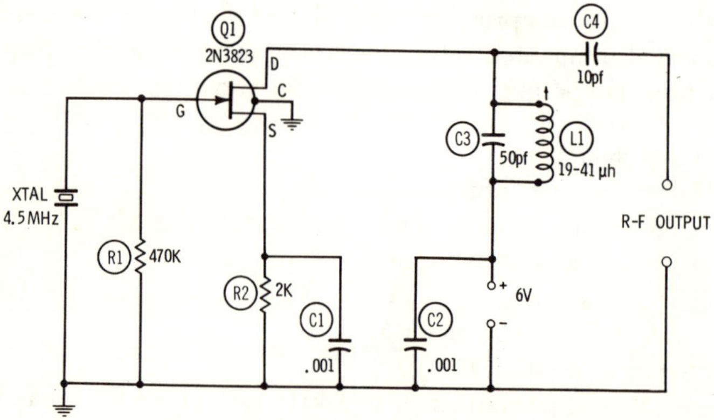

Fig. 7-8. Television sound-marker generator (4.5-MHz).

the slug-tuned tank coil, L1 (Miller No. 42A335CBI, or equivalent) for maximum output, as indicated by a communications receiver or an r-f voltmeter connected to the R-F OUTPUT terminals.

The open-circuit signal amplitude at the R-F OUTPUT terminals is 4.0 volts rms when the d-c supply is 6 volts at 1.4 ma. The signal strength may be continuously adjusted by making C4 a 50-pf midget variable capacitor.

Resistors R1 and R2 are 1/2 watt. Capacitors C1, C2, and C4 are mica, and capacitor C3 is silvered mica. The fourth pigtail of the 2N3823 is internally connected to the metal case of this FET and must be grounded as shown.

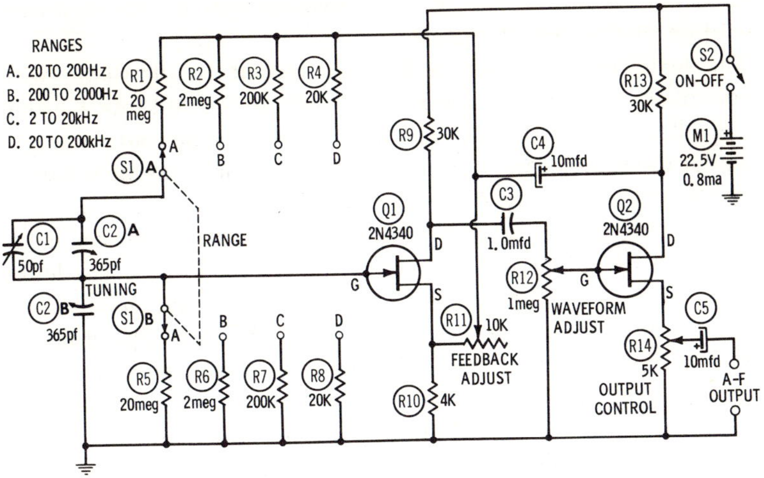

Fig. 7-9. Variable-frequency audio oscillator.

VARIABLE-FREQUENCY AUDIO OSCILLATOR

Fig. 7-9 shows the circuit of a variable-frequency r-c–tuned audio oscillator employing two 2N4340 FET's. Using the Wien-bridge frequency-determining network, this instrument covers 20 Hz to 200 kHz in four ranges: 20-200 Hz, 200-2000 Hz, 2-20 kHz, and 20-200 kHz. Tuning in each range is accomplished with a dual 365-pf variable capacitor (C2), and ranges are changed by switching identical resistors in pairs (R1-R5, R2-R6, R3-R7, and R4-R8). These resistances must be accurate to at least 1 percent.

In this circuit there are two feedback paths, both through capacitor C4. One is through the bandswitched Wien bridge to the gate of FET Q1, the other is through FEEDBACK ADJUST rheostat R11 to the source of Q1. At the resonant frequency of the Wien bridge, the feedback to the Q1 gate is positive and creates oscillation. The feedback through R11, however, is negative and, since it cancels the oscillation on all frequencies except the bridge resonant frequency, is responsible for the very low distortion of this type of circuit.

The amplitude of the negative feedback is adjustable with rheostat R11, and the amplitude of the driving voltage presented to the feedback/output amplifier (Q2) is adjustable with potentiometer

R12. Optimum adjustment of these two controls results in lowest distortion of the output signal.

The low-impedance, source-follower output is adjustable with potentiometer R14 from approximately zero to 0.4 volt rms. Current drain is 0.8 ma from the 22.5-volt battery, M1. Higher output amplitude (at the same current drain for the oscillator) may be obtained with an additional output-amplifier stage.

The tuning capacitor (C2) must be insulated from the metal chassis and should be enclosed in a grounded shield box. The lower section of this capacitor is grounded, but the upper section floats and must be compensated with the 50-pf trimmer capacitor, C1. Solid construction must be employed throughout, and all wiring must be short, direct, and rigid.

The oscillator may be calibrated in the conventional manner against any available standard audio frequency, using an oscilloscope set up for Lissajous figures. Another method makes use of a calibrated audio-frequency meter (see an earlier section of this chapter) connected to the A-F OUTPUT terminals. With switch S1 set to the proper range, capacitor C2 is adjusted to resonance with the standard frequency (or to the desired frequency, as indicated by the audio-frequency meter). The dial of this tuning capacitor then is set to that frequency. Trimmer capacitor C1 is adjusted for continuous oscillation throughout the range.

Rheostat R11 is adjusted for constant output over the entire range of the oscillator. as indicated by an electronic a-c voltmeter or oscilloscope connected to the A-F OUTPUT terminals. Potentiometer R12 is adjusted for minimum distortion (as indicated by an oscilloscope or distortion meter connected to the A-F OUTPUT terminals). After adjustment, these two variable resistors need not be touched again until the instrument is routinely rechecked.

All fixed resistors are 1 watt. Capacitors C4 and C5 are 50-dcwv electrolytics, and capacitor C3 is a 200-volt metallized paper tubular. Rheostat R11 and potentiometer R12 have slotted shafts for screwdriver adjustment and are mounted safely inside the instrument case to protect them from tampering. Trimmer capacitor C1 is also inside the case, being securely fastened to the frame of tuning capacitor C2. The gate electrode of the 2N4340 is internally connected to the metal case of this FET, so both 2N4340's must be kept clear of contact with each other, other components, wiring, or chassis.

DIP OSCILLATOR

The tube-type grid-dip oscillator is a well-known and useful instrument which needs no theoretical explanation nor a recitation of its merits here. Previously, some transistorized versions of this device, while offering freedom from the power line, have been less satisfying than the tube type, since the bipolar transistor reduced sensitivity and selectivity of the circuit. With an FET, however, the dip oscillator behaves just as it does with a tube, because the gate-to-ground impedance of the FET is comparable to the grid-to-ground impedance of the tube. Response is sharp and fast, and the indicating meter need not be connected in some special part of the circuit different from the grid circuit of the tube.

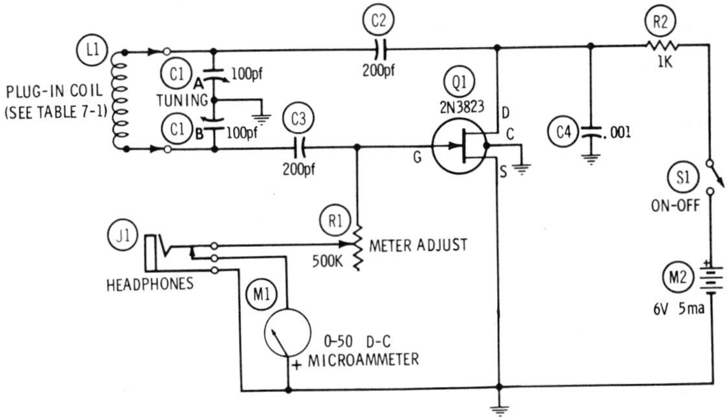

Fig. 7-10. Dip oscillator.

Fig. 7-10 shows the circuit of a dip oscillator, based on a 2N3823 FET. With six plug-in coils, this instrument covers the range 1.1 to 95 MHz: (A) 1.1-2.5 MHz, (B) 2.5-5 MHz, (C) 5-11 MHz, (D) 10-25 MHz, (E) 20-45 MHz, and (F) 40-95 MHz. The circuit is that of a Colpitts oscillator tuned with a dual 100-pf variable capacitor, C1.

Dip is indicated by the 0-50 d-c microammeter, M1, in the gate-to-ground leg, and the meter sensitivity is adjustable with rheostat R1. Downward deflection is sharp when the tank (L1-C1) is tuned to the resonant frequency of an external l-c circuit to which the tank is loosely coupled.

Like other dip oscillators, this one is useful for checking signals as well as dead l-c circuits. A closed-circuit jack, J1, is provided for high-resistance magnetic headphones used for this purpose. When the headphones are plugged in, the microammeter is automatically disconnected from the circuit.

The instrument may be calibrated in either of two ways: (1) Loosely couple L1 to a succession of dead l-c tank circuits whose frequencies are accurately known, tune C1 successively for dip at the various frequencies, and mark the dial of this variable capacitor correspondingly, or (2) With headphones plugged into jack J1, tune in a number of accurately known signals, adjusting C1 for zero beat at each frequency and marking the dial of the tuning capacitor correspondingly.

For stability, the instrument must be solidly constructed, and the shortest, most direct, and most rigid wiring practicable must be used. All coils are tightly wound on 1-inch-diameter plug-in forms (see Table 7-1 for coil-winding data). The number of turns in the coils may need some adjustment to compensate for stray capacitance in an individual instrument and for the particular gate-to-source capacitance of an individual FET. Resistor R2 is 1/2 watt, and capacitors C2, C3, and C4 are silvered mica. The fourth pigtail of the 2N3823 is internally connected to the metal case of this FET and must be grounded as shown. Current drain, from the 6-volt battery, M2, is 5 ma.

The steady deflection of the microammeter does not have the same value on all frequency ranges and may even change to a higher or lower value as the instrument is tuned through a single

Table 7-1. Coil-Winding Data*

Coil	Frequency Range	Specifications
A	1.1-2.5 MHz	150 turns No. 32 enameled wire, closewound
B	2.5-5 MHz	77 turns No. 28 enameled wire, closewound
C	5-11 MHz	35 turns No. 22 enameled wire, closewound
D	10-25 MHz	17 turns No. 22 enameled wire, spaced to winding length of 1 inch
E	20-45 MHz	8.5 turns No. 22 enameled wire, spaced to winding length of 1 inch
F	40-95 MHz	4½ turns No. 22 enameled wire, spaced to winding length of 1 inch

*All coils wound on Millen No. 45004 1-inch, 4-pin coil forms, or equivalent (use two of the pins).

range. Adjustment of rheostat R1, however, enables the operator to set the pointer of M1 to a desirable point on the scale before tuning for dip. This rheostat also provides some degree of volume control when the instrument is used with headphones.

A-F MILLIAMMETER

Fig. 7-11 shows the circuit of an audio-frequency milliammeter, based on a single 2N4868 FET (Q1), which covers the current range 0 to 500 ma in seven full-scale ranges: 0.5, 1, 5, 10, 50, 100, and 500 milliamperes. A 2-pole, 7-position, rotary selector switch (S1) switches in the appropriate values of shunt resistance (R1 to R7). The double-pole switching guarantees that the circuit is never

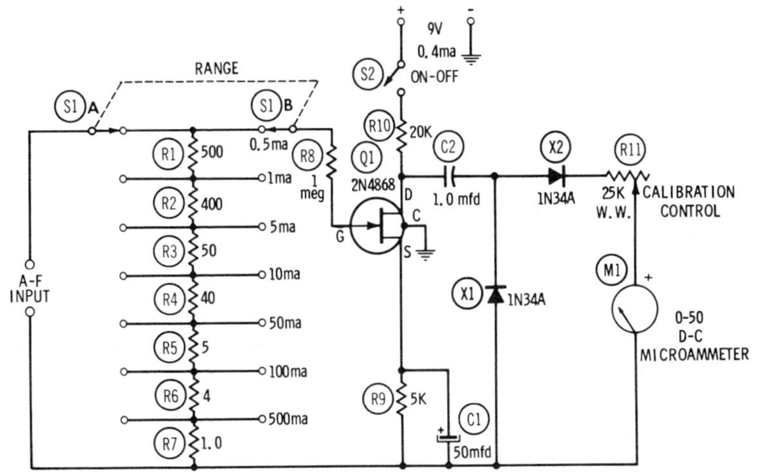

Fig. 7-11. A-f milliammeter.

exposed to excessive current between switch positions. Linearity of response is within 5 percent of full scale. Frequency response, referred to 1 kHz, is down 3.5 db at 50 Hz and 2 db at 50 kHz. Current drain is 0.4 ma at 9 vdc.

Input-resistance values are: 0.5 ma, 1000 ohms; 1 ma, 500 ohms; 5 ma, 100 ohms; 10 ma, 50 ohms; 50 ma, 10 ohms; 100 ma, 5 ohms; and 500 ma, 1 ohm.

Like the a-f millivoltmeters described earlier in this chapter, this circuit uses the conventional amplifier-rectifier arrangement: the signal is developed by current through the selected shunt resistors and is first amplified by the FET and then rectified by the two germanium diodes (X1, X2). The d-c output of the diodes is more than sufficient to drive microammeter M1 to full scale. The meter deflection is proportional to the average value of the input a-f current, but the meter may be calibrated to read rms current on a sine-wave basis.

No zero adjustment is needed. Rheostat R11, the CALIBRATION control, is mounted safely inside the instrument case to protect it from being turned by accident after the instrument has been completely calibrated.

The 0 to 50 d-c microampere scale of meter M1 reads directly on the 50-ma range, and is easily read on the 0.5-, 5-, and 500-ma ranges by mentally shifting the decimal point as required. On the 1-, 10-, and 100-ma ranges, however, the reading must be mentally multiplied by 2 or some multiple of 2 (thus, by 0.02 for the 1-ma, by 0.2 for the 10-ma, or by 2 for the 100-ma range of the a-f milliammeter).

Calibrate the instrument in the following manner:

1. Set switch S1 to the 50-ma range.
2. Set rheostat R11 to full resistance.
3. Close switch S2.
4. Connect an accurately known 50-ma a-c source to the A-F INPUT terminals (any frequency between 60 Hz and 1 kHz).
5. Adjust rheostat R11 for exact full-scale deflection of meter M1. This rheostat then need not be touched again until the instrument is routinely recalibrated.

The instrument is now in calibration on all ranges if accurate values have been used for resistors R1 through R7.

All fixed resistors are 1 watt. (The range resistors—R1 to R7—must be accurate to at least 1 percent.) Capacitor C1 is a 25-dcwv electrolytic; C2 is a 100-volt plastic unit. Rheostat R11 is a wire-wound unit provided with a slotted shaft for screwdriver adjustment.

For stability, all wiring must be as short, rigid, and direct as practicable; and the fourth pigtail of the 2N4868, which is internally connected to the metal case of this FET, must be grounded as shown, for shielding.

INDUCTANCE CHECKER

Any inductance between 60 μh and 60,000 hy may be checked by the resonance method, using the circuit shown in Fig. 7-12 with a variable-frequency audio oscillator. With the unknown inductance connected to terminals X-X, the oscillator is tuned for peak deflection of meter M1, whereupon the inductance is determined (in terms of the .01-mfd capacitance, C3) by calculation or by reference to the graph in Fig. 7-13.

In this circuit (Fig. 7-12), the U183 FET (Q1) serves as a source follower between the oscillator and the measuring circuit. The advantages of this arrangement are (a) the high input impedance of the resultant circuit reduces oscillator loading to virtually zero, and (b) the follower isolates the measuring circuit (M1-X1-C3-L_x) from the signal source.

Potentiometer R1 allows adjustment of the gain for exact full-scale resonant deflection of meter M1. This is an added convenience even when the audio oscillator has its own output control. With R1 set for maximum gain, an oscillator output of 1.5 volts rms will drive meter M1 to full scale with most inductors which will be connected to terminals X-X. D-c microammeter M1 is made to respond to alternating current simply by connecting the 1N34A germanium diode (X1) in parallel with it. Current drain of the circuit is 3.3 ma at 9 vdc.

Resistor R2 is 1/2 watt. Capacitor C1 is a 100-volt plastic unit; C2 is a 25-dcwv electrolytic. For stability and accuracy, C3 must be a silvered mica capacitor as close as practicable to the rated .01 mfd, since it is with reference to this capacitance that the unknown inductance is determined.

The checker must be built mechanically solid, and all wiring must be as short, rigid, and direct as possible. A fourth pigtail of the U183 is internally connected to the metal case of this FET and must be grounded as shown, for shielding.

To use the checker:

1. Connect a variable-frequency audio oscillator (tuning range: 20 to 200,000 Hz) to the OSCILLATOR INPUT terminals.
2. Connect the unknown inductance (L_x) to terminals X-X.
3. Close switch S1, and switch on the oscillator.

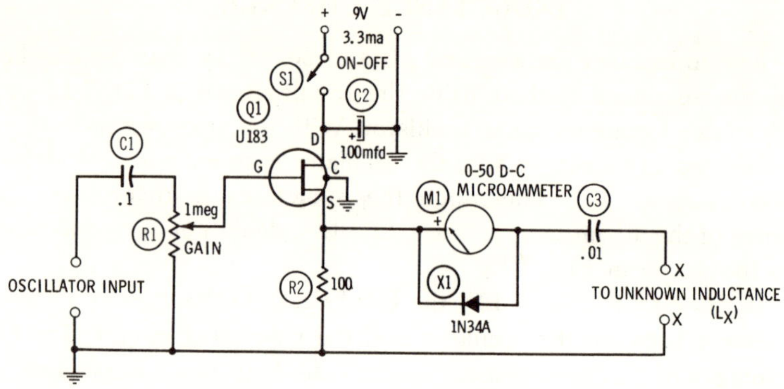

Fig. 7-12. Inductance checker.

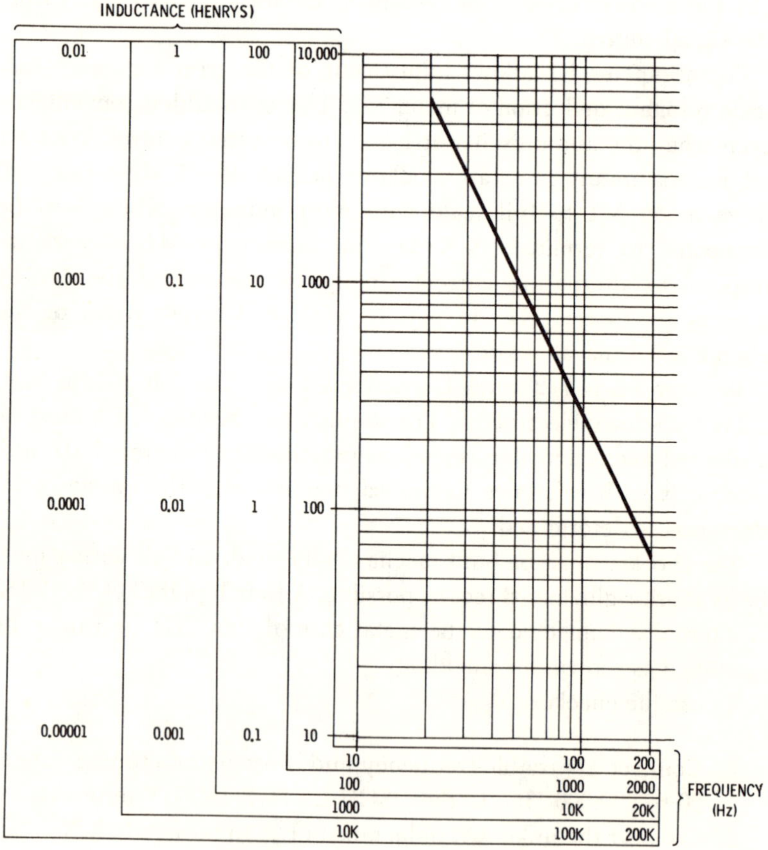

Fig. 7-13. Inductance chart.

4. Starting at the botom of the lowest frequency range, tune the oscillator slowly throughout its ranges until a sharp deflection of meter M1 occurs. Adjust the oscillator output control and gain-control potentiometer R1 to bring this deflection close to full scale of the meter.
5. At this point of peak deflection, read the frequency from the oscillator dial.
6. For close results, calculate the inductance:

$$L_x = \frac{1}{(395\, f^2 \times 10^{-9})}$$

where,

L_x is the unknown inductance, in henrys,

f is the oscillator frequency, in hertz.

7. For approximate results, eliminating calculations, use the chart given in Fig. 7-13.

SUBSTITUTION-TYPE CAPACITANCE METER

Small capacitances are accurately measured by means of the substitution process, since this method of measurement automatically compensates for stray capacitances in the test setup. It is advantageous to measure small capacitances (0.1 to 400 pf) at radio frequencies. Fig. 7-14 shows the circuit of a substitution-type capacitance checker operated at 1 MHz.

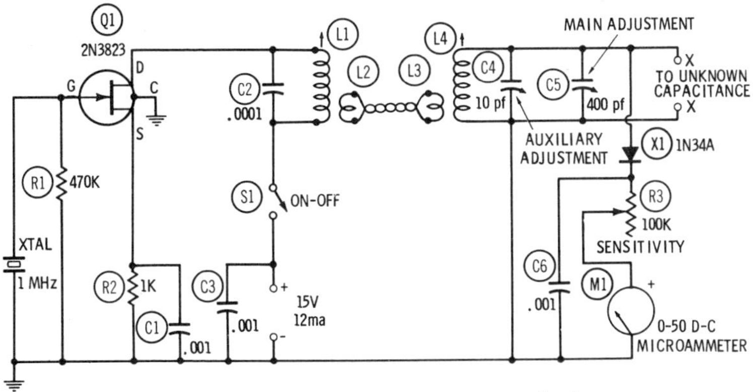

Fig. 7-14. Substitution-type capacitance checker.

This arrangement employs a 1-MHz crystal oscillator based on a 2N3823 FET (Q1). The fix-tuned oscillator tank (L1-C2) is link-coupled to the resonant measuring circuit, which is composed of inductor L4, trimmer C4, variable capacitor C5, and unknown capacitance C_x connected to terminals X-X. A simple r-f voltmeter (consisting of germanium diode X1, rheostat R3, microammeter M1, and r-f bypass capacitor C6) is connected across the measuring circuit as a resonance indicator. Both C4 and C5 are provided with calibrated dials reading directly in picofarads. The oscillator coil (L1) and the measuring-circuit coil (L4) are slug-tuned units (L1, adjusted to 250 μh, Miller No. 20A224RBI; L4, adjusted to 60 μh, Miller No. 41A685CBI). L2 and L3 are link-coupling coils which must be provided by the builder; each consists of 2 turns of insulated hookup wire closewound near the bottom ends of L1 and L4, respectively.

All fixed resistors are 1/2 watt. Capacitors C1, C3, and C6 are mica; C2 is silvered mica. All wiring must be short, direct, and rigid. Moreover, the wiring in the measuring circuit (i.e., between L4, C4, C5, and terminals X-X) must be straight, solid No. 14 bus. A fourth pigtail of the 2N3823 is internally connected to the metal case of this FET and must be grounded as shown, for shielding. Total current drain is 12 ma at 15 vdc.

After the assembly has been completed and the wiring verified, the crystal oscillator must be initially adjusted:

1. Connect an r-f vtvm between ground and the top of L1.
2. Close switch S1.
3. Adjust the L1 slug for peak deflection of the vtvm. This slug need not be touched again except during routine recalibration of the instrument.

Next, initially adjust the measuring circuit:

1. Set variable capacitor C4 to half capacitance and C5 to maximum capacitance.
2. Set rheostat R3 to maximum resistance.
3. Close switch S1.
4. Adjust the L4 slug for peak deflection of meter M1, reducing the resistance of R3 if necessary to ensure a readable deflec-

tion. This slug need not be touched again except during routine recalibration of the instrument.

To use the checker:

1. Close switch S1, and resonate the circuit by setting C5 to maximum capacitance and trimming with C4 for peak deflection of meter M1. Thenceforth, do not touch C4.
2. Record the capacitance setting of C5 as C_a.
3. Connect the unknown capacitance (C_x) to terminals X-X by the shortest possible leads. This will detune the measuring circuit, and the meter deflection will either decrease or disappear.
4. Retune C5 to restore resonance, as indicated by peak deflection of M1.
5. Record this new capacitance setting of C5 as C_b.
6. Calculate the unknown capacitance:

$$C_x = C_a - C_b$$

When long leads *must* be used to connect the unknown capacitor to the instrument, install the leads first (in the same position they will take when the capacitor later is connected to them) and then perform Steps 1 and 2. This automatically compensates for the lead capacitance. Then connect the unknown capacitor to the far end of the leads and perform Steps 3 to 6 in the same manner described above.

When very small capacitances (0.1 to 10 pf) are measured, all readings (C_a and C_b) should be taken from the dial of C4, and capacitor C5 should be used only to trim the circuit to resonance in Step 1. This step then consists of the following procedure:

1. Set C4 to maximum capacitance. Close switch S1, and trim with C5 for peak deflection of meter M1. Record this setting of C4 as C_a, and thenceforth do not touch C5.

IMPEDANCE METER

Fig. 7-15 shows the circuit of an impedance meter which employs a widely used principle of measurement: an a-f voltage is applied to an unknown impedance (Z) in series with a calibrated variable resistor (R), and the voltage drops (E_z and E_r) across Z and R, respectively, are checked separately with an electronic voltmeter while R is varied. When the voltage drops are equal (i.e., $E_z = E_r$), Z = R and may be read directly from the calibrated dial of the variable resistor.

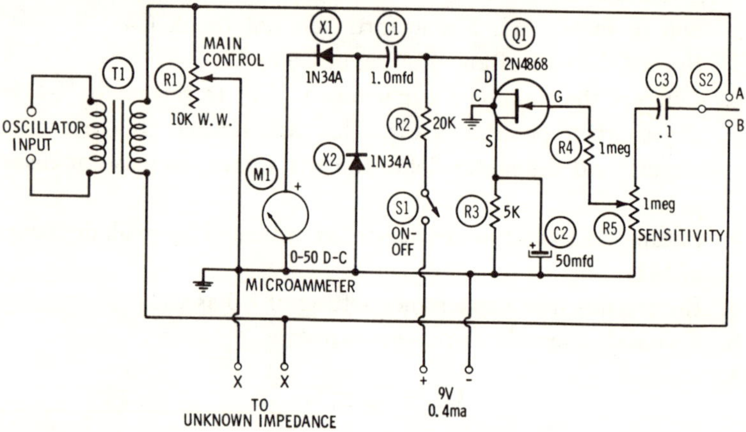

Fig. 7-15. Impedance meter.

In Fig. 7-15, rheostat R1 is the calibrated variable resistor (0 to 10,000 ohms) which is in series with the unknown impedance connected to terminals X-X. A signal voltage, derived from an external audio oscillator, is applied to this series combination through input transformer T1. The electronic voltmeter is composed of an a-f amplifier, based on a 2N4868 FET (Q1), followed by a 2-diode rectifier (X1, X2) and a d-c microammeter (M1). When switch S2 is in Position A, this voltmeter is connected across variable resistor R1 and reads E_r; when S1 is in Position B, the voltmeter is connected across the unknown impedance and reads E_z. The meter needs no voltage calibration, since it serves only to indicate reference points; the d-c scale of microammeter M1 may be used directly for this purpose. (It is only necessary to adjust R1 so that the *deflection* is the same for each position of switch S2; the magnitude of the deflection is inconsequential.) The input potentiometer (R5) of the voltmeter circuit allows the meter to be adjusted for deflection near full scale to increase accuracy. Current drain of the circuit is 0.4 ma at 9 vdc.

All fixed resistors are 1/2 watt. Rheostat R1 is wirewound and must have a dial reading directly in ohms. Capacitors C1 and C3 are 100-volt plastic units; C2 is a 25-dcwv electrolytic. Transformer T1 may be any convenient unit having a 1:1, 2:1, or 3:1 turns ratio, and preferably should be shielded.

All wiring must be as short, rigid, and direct as practicable. Overall shielding is desirable but not mandatory. A fourth pigtail of the 2N4868 is internally connected to the metal case of this FET and must be grounded as shown, for shielding.

Use this instrument in the following manner:

1. Connect an audio oscillator to the OSCILLATOR INPUT terminals (common test frequencies are 400, 500, and 1000 Hz), switch it on, and set its output to a level which will not saturate transformer T1.
2. Connect the unknown impedance (Z_x) to terminals X-X.
3. Set potentiometer R5 for maximum sensitivity.
4. Close switch S1.
5. Throw switch S2 to Position A, and adjust rheostat R1 and potentiometer R5 for a deflection near full scale of meter M1. Note the meter reading.
6. Throw switch S2 to Position B and note the meter reading.
7. Adjust R1 to make the deflection the same as it was in Step 5.
8. Throw switch S2 back to Position A, note the deflection and adjust R1 again, as in Step 7.
9. Continue this procedure until there is no change in meter reading as S2 is thrown back and forth between A and B.
10. At this point, read the value of the unknown impedance from the ohms dial of rheostat R1.

HETERODYNE FREQUENCY METER

By means of the zero-beat method radio frequencies far into the spectrum may be spotted and measured with a calibrated low-frequency r-f oscillator. This is made possible by zero beating between the unknown frequencies and harmonics of the calibrated oscillator. When a mixer and calibrated oscillator are combined for this purpose in one unit, the result is a *heterodyne frequency meter.*

Fig. 7-16 shows the circuit of a heterodyne frequency meter consisting of a 1 to 2-MHz oscillator (Q1), an untuned mixer (X1), and an a-f (beat-note) amplifier (Q2). The oscillator employs a 2N3823 FET; the mixer, a 1N34A germanium diode; and the a-f amplifier, a TIS14 FET.

The oscillator is of the Hartley type and is continuously tunable over the range of 1 to 2 MHz. The oscillator tank comprises coil L1, 365-pf tuning capacitor C4, and 50-pf trimmer capacitor C3. Coil L1 consists of 65 turns of No. 28 enameled wire closewound on a 1-inch–diameter form, with a tap taken at the 20th turn from the ground end. L2 is a coupling coil consisting of 10 turns of No. 28 enameled wire closewound around the center of L1 and insulated from the latter.

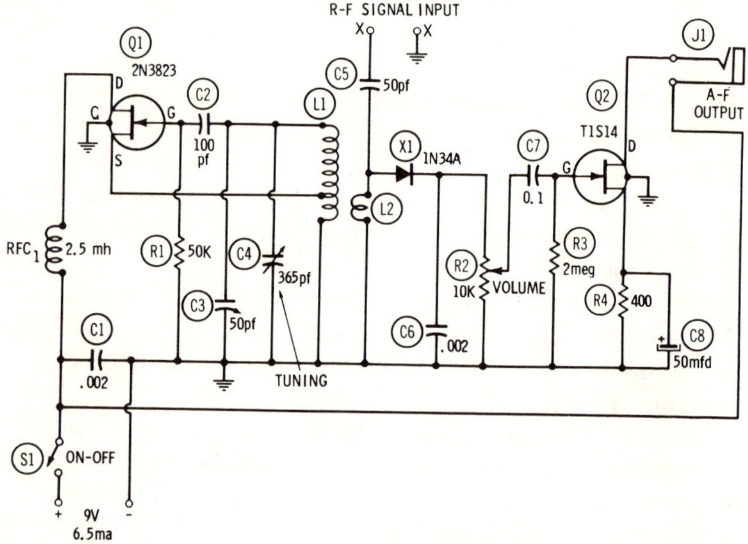

Fig. 7-16. Heterodyne frequency meter.

The oscillator output is coupled, through L2, into the diode mixer simultaneously with the unknown r-f signal (coupled through capacitor C5). A beat note between the two signals is generated in the mixer and is transmitted to the a-f amplifier through coupling capacitor C7. The latter also serves to block the diode d-c component from the TIS14 gate. Magnetic headphones, plugged into jack J1, complete the TIS14 drain circuit and make the beat note audible. Other beat-note indicators, such as a vtvm, oscilloscope, magic-eye tube, or audio-frequency meter, may be used, provided J1 is shunted by a 2000-ohm resistor to complete the TIS14 d-c circuit.

The instrument draws 6.5 ma at 9 vdc. This value is constant for zero signal and maximum signal.

All fixed resistors are 1/2 watt. Capacitors C1, C5, and C6 are mica; C2 is silvered mica. C7 is a 100-volt plastic unit, and C8 a 25-dcwv electrolytic.

The oscillator tuning unit (L1-C3-C4) must be stoutly built, and the entire oscillator stage should be enclosed in a grounded shield box. All wiring must be as short, direct, and rigid as practicable. A fourth pigtail of both the 2N3823 and the TIS14 is internally connected to the metal case of these FET's and must be grounded as shown, for shielding.

Main tuning capacitor C4 must be provided with a dial calibrated to read directly in frequency—1000 to 2000 kHz. One way to calibrate this dial is to feed an accurate, unmodulated r-f signal-generator output into the R-F SIGNAL INPUT terminals, set C3 to midcapacitance, set the generator successively to as many frequencies as practicable between 1000 and 2000 kHz, and at each frequency set C4 to zero beat (as indicated by headphone signal) and inscribe the C4 dial with the corresponding frequency. This calibration then may be checked more precisely by replacing the generator with a 100-kHz crystal oscillator, which will set up an accurate spot frequency each 100 kHz across the C4 dial. To standardize the oscillator in the future, it is only necessary to feed in the 100-kHz oscillator, set the C4 dial to its 1000-kHz point, and adjust trimmer C3 for zero beat. Since C3 needs no constant adjustment, it should be mounted inside the instrument case, where it will be safe from disturbance, and should be provided with a slotted shaft for screwdriver adjustment.

On second harmonics, the 1 to 2-MHz C4 dial covers 2 to 4 MHz, on twentieth harmonics 20 to 40 MHz, and so on. Similarly, on the second "subharmonic" the dial covers 500 to 1000 kHz.

HARMONIC-DISTORTION METER

A harmonic-distortion meter is convenient for measuring the total distortion of an audio amplifier, a component, or a network. A pure sine-wave signal is applied to the input of the device under test and

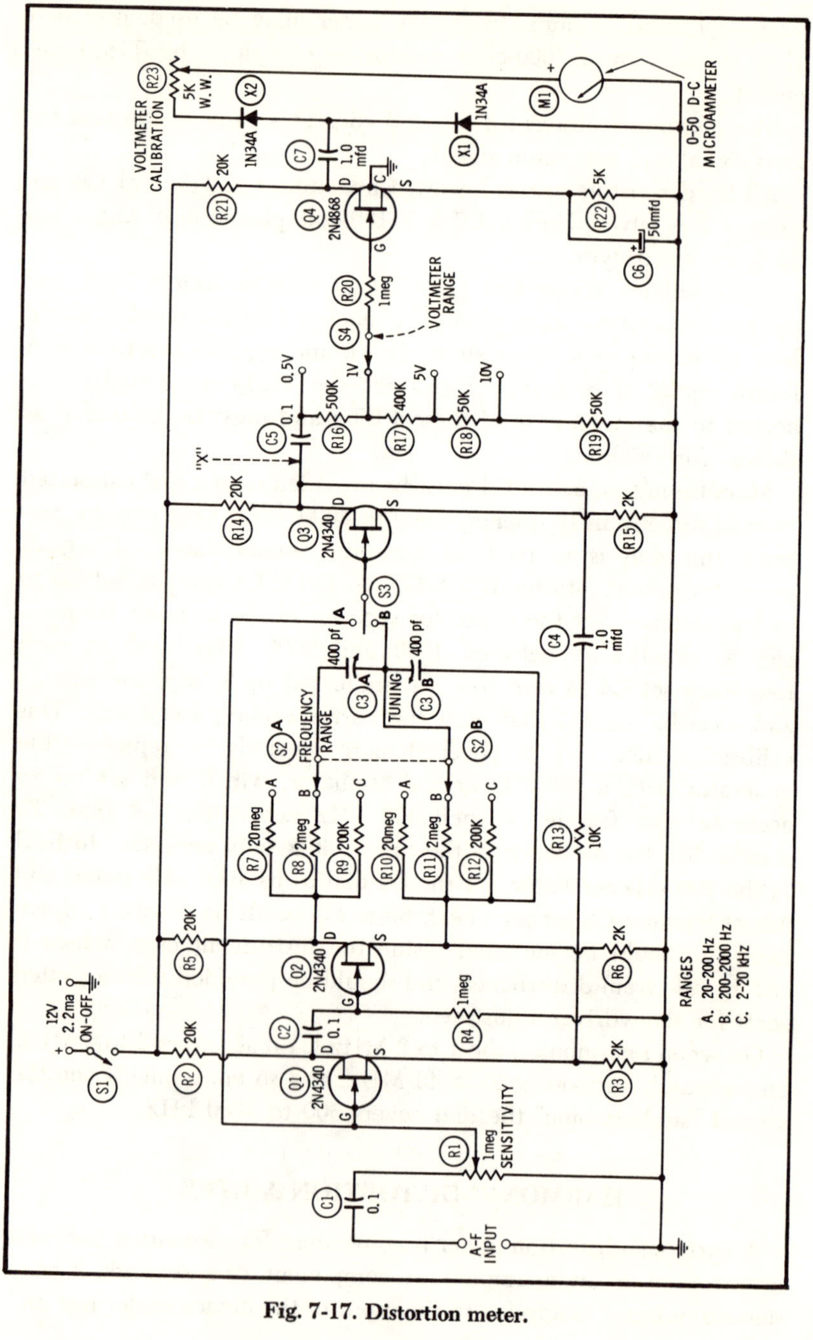

Fig. 7-17. Distortion meter.

the distortion meter is connected to the output. The meter is tuned to suppress the fundamental frequency of the test signal. The self-contained electronic-voltmeter section of the instrument then is used to sample the output-signal voltage before and after the suppression. The first voltage contains the fundamental and harmonics, whereas the second one contains only the harmonics (distortion components introduced by the device under test). The total distortion therefore is the ratio of the second voltage to the first, and can be expressed either as a decimal or as a percentage.

Fig. 7-17 shows the circuit of a complete distortion meter employing these principles. This instrument covers the frequency range of 20 Hz to 20 kHz in three bands: (A) 20-200 Hz, (B) 200-2000 Hz, and (C) 2-20 kHz. Ranges are selected by double-pole, 3-position switch S2. The filter circuit is composed of range resistors R7 to R12 and dual 400-pf tuning capacitor C3. Within the amplifier comprised by FET's Q1, Q2, and Q3, this r-c network nulls sharply at the frequency to which it is tuned, and allows signals at other frequencies to pass relatively untouched in amplitude. The r-c network alone tunes broadly, but its response is sharpened by overall negative feedback around the amplifier through capacitor C4 and resistor R13. The voltmeter section comprises FET Q4, input range selector S4 and R16-R19, and rectifier-type microammeter (M1-X1-X2). Total current drain is 2.2 ma at 12 vdc.

The amplifier section is based on three 2N4340 FET's (Q1, Q2, Q3). This section is followed by an a-f voltmeter based on a single 2N4868 FET (Q4). When switch S3 is in Position A, the voltmeter checks the signal voltage before the filter; and when S3 is in Position B, the meter checks the signal voltage after the filter.

All fixed resistors are 1 watt. Resistors R2, R3, R5, R6 to R12, and R14 to R19 must be rated at 1 percent. In an individual layout, R13 may need adjustment for sharpest null response of the circuit. Capacitors C1, C2, C4, C5, and C7 are 100-volt plastic units; C6 is a 25-dcwv electrolytic. Each section of the dual 400-pf tuning capacitor (C3) has a built-in trimmer for initial adjustment.

After the instrument has been completed and the wiring verified, first the voltmeter section should be calibrated:

1. Set potentiometer R1 for maximum sensitivity.
2. Set switch S3 to Position A.

3. Set switch S4 to its 5-volt position.
4. Connect an accurately known 5-volt (rms) source to the A-F INPUT terminals.
5. Close switch S1.
6. Adjust rheostat R23 for exact full-scale deflection of microammeter M1.

The voltmeter now is calibrated on all ranges if resistors R16 to R19 are accurate. (R23 need not be touched again except during routine recalibration).

Next, the filter (tuner) should be adjusted:

1. Set potentiometer R1 for maximum sensitivity.
2. Set switch S2 to Position A.
3. Set switch S3 to Position B.
4. Set switch S4 to its 1-volt position.
5. Connect a high-grade audio oscillator to the A-F INPUT terminals and set it to 200 Hz at 1-volt rms output.
6. Set tuning capacitor C3 to minimum capacitance.
7. Close switch S1.
8. By adjusting the trimmers on C3A and C3B, tune the circuit for full dip (null) of the M1 deflection. (Readjust R1 and S4, as required, for readable meter response.)
9. Tune C3 successively to various other frequencies down to 20 Hz, inscribing the C3 dial accordingly. The filter now will automatically be calibrated on the two other bands (B and C) if resistors R7 to R12 are accurate.

Use of the instrument is straightforward:

1. Connect a high-grade sinewave audio oscillator to the input of the device under test.
2. Connect the distortion meter to the output of the device.
3. Set switch S2 to the appropriate frequency range.
4. Close switch S1.
5. Set switch S3 to Position A.
6. With switch S4 in the appropriate voltage position, adjust potentiometer R1 for full-scale deflection of meter M1.
7. Record this voltage as E_1.

8. Set switch S3 to Position B.
9. Tune capacitor C3 for null, and set S4 to successively lower ranges for accurate reading of the voltage at null.
10. Record the residual null voltage as E_2.
11. Calculate the distortion percentage:

$$D\ \% = 100\ \frac{E_2}{E_1}$$

Under certain conditions, the instrument may be direct reading in distortion percentage, requiring no calculations. Thus, if E_1 is 1 volt, this point may be regarded as 100 percent. The distortion percentages at null then may be read directly; e.g.. 0.01 volt = 1 percent, 0.1 volt = 10 percent, and so on.

A-F R-F SIGNAL TRACER

The high input impedance of the FET makes possible for the first time a transistorized signal tracer which is fully competitive with the tube-type tracer. Basically, the signal tracer is a high-gain a-f amplifier which drives a visual signal-strength indicator and a speaker, and to which may be attached either a simple test probe for a-f input or a demodulator probe for a-m r-f input.

Fig. 7-18 shows the circuit of a signal tracer employing two field-effect transistors (Q1, Q2) and three NPN bipolar transistors (Q3, Q4, Q5). High sensitivity is provided by the front end, which consists of an r-c–coupled audio amplifier based on one 2N4868 FET (Q1) and one 2N4340 FET (Q2). The a-f probe or r-f probe, as required, is plugged into this amplifier through concentric jack J1. The front-end amplifier drives meter M1 through germanium diodes X2 and X3. This provides the visual indication of signal strength.

For aural indication of signal strength, a speaker is driven by the power amplifier consisting of a driver stage (based on a 2N169A transistor, Q3) and a push-pull class-B output stage (based on two 2N169A transistors, Q4 and Q5). The latter stage delivers 100 milliwatts to the speaker.

Switch S1 allows the speaker section to be turned off when the silent, visual indication provided by the meter is sufficient. At 9 vdc, the signal tracer draws a zero-signal current of 11 ma whether the speaker is on or off, and a maximum-signal current of 25 ma when the speaker is on and 11 ma when it is off.

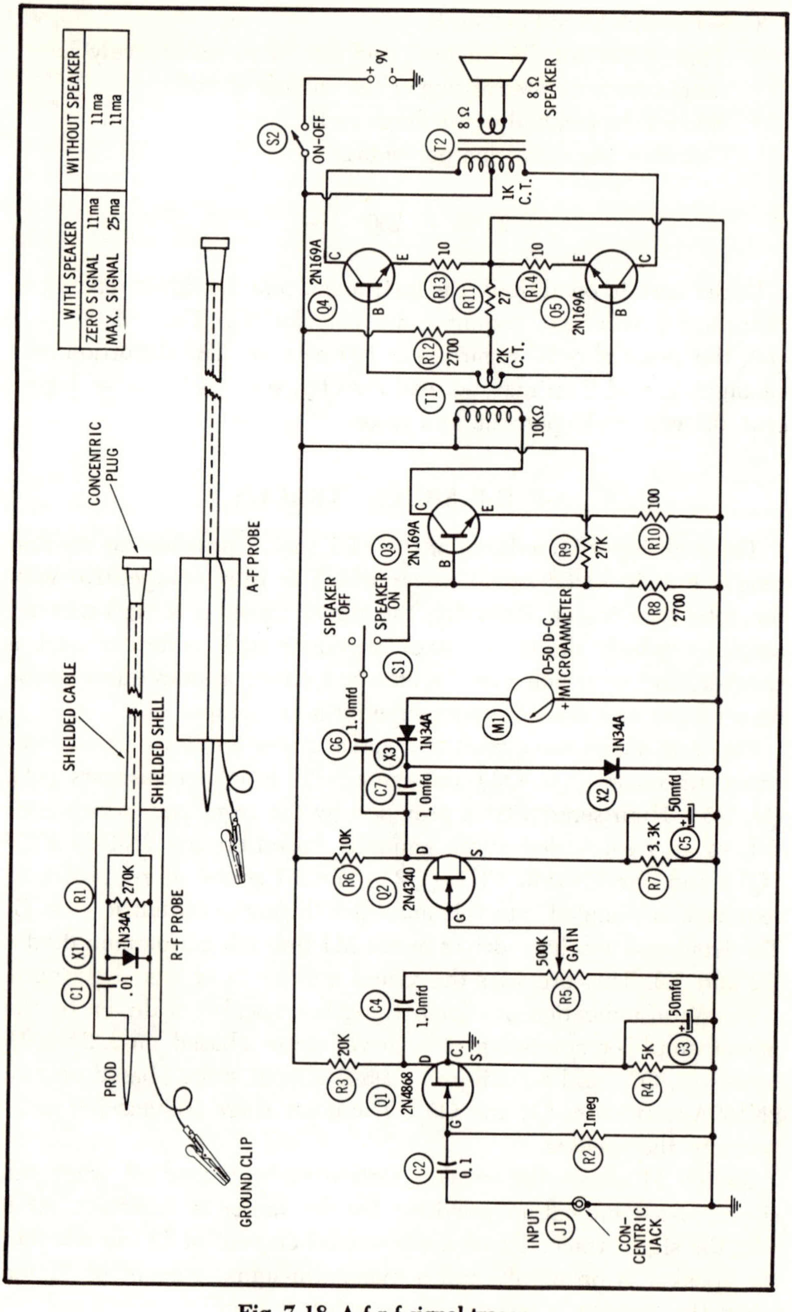

Fig. 7-18. A-f r-f signal tracer.

All fixed resistors are 1/2 watt. Capacitor C1 is a 500-volt ceramic unit, while C2, C4, C6, and C7 are 100-volt plastic or paper units. C3 and C5 are 25-dcwv electrolytics.

The layout is not critical; however, all wiring must be as short and direct as possible, to minimize stray coupling and interference. For shielding, the entire unit must be enclosed in a metal box.

The instrument needs no calibration; since response of the meter section is linear, the readings of microammeter M1 may be taken directly as indications of signal strength, as long as the setting of gain-control potentiometer R5 is undisturbed during a measurement.

SENSITIVE LIGHT METER

The P-102 photo-FET is the basis of the sensitive light-meter circuit shown in Fig. 7-19. In this instrument, full-scale deflection of the 0 to 100 d-c microammeter (M1) is obtained with only 5 foot-candles of illumination, when R1 is 10 megohms. Other ranges may be obtained by changing the value of this resistor—the lower the resistance of R1, the higher the range.

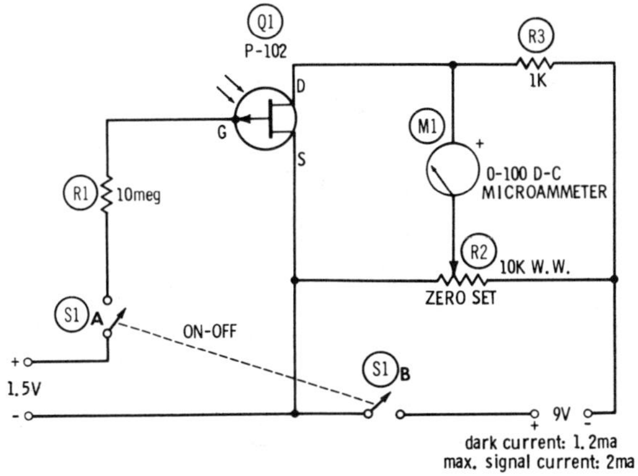

Fig. 7-19. Sensitive light meter.

Since there is some static drain current through the P-102, potentiometer R2 has been provided for setting the meter to zero. The operation consists simply of darkening the P-102 and adjusting R2 for zero deflection of meter M1.

Current drain of the circuit is 1.2 ma at 9 vdc when the P-102 is darkened, and 2 ma when the light input deflects the meter to full scale.

Layout of the instrument is not critical. However, the P-102 has a lens in its nose, so this photo-FET must be mounted so as to point toward the light source. And since the gate electrode is internally connected to the case of the P-102, the FET must be mounted clear of wiring, chassis, and other components. Both fixed resistors are 1/2 watt, and potentiometer R2 is a wirewound unit.

GEIGER COUNTER

Fig. 7-20 shows the circuit of a simple Geiger counter based on one 2N4338 FET (Q1) and one 2N4868 FET (Q2). Basically, this instrument is a high-gain r-c–coupled a-f amplifier followed by an a-c meter circuit (coupling capacitor C4, germanium diodes X1 and X2, microammeter M1, and high-capacitance meter capacitor C6).

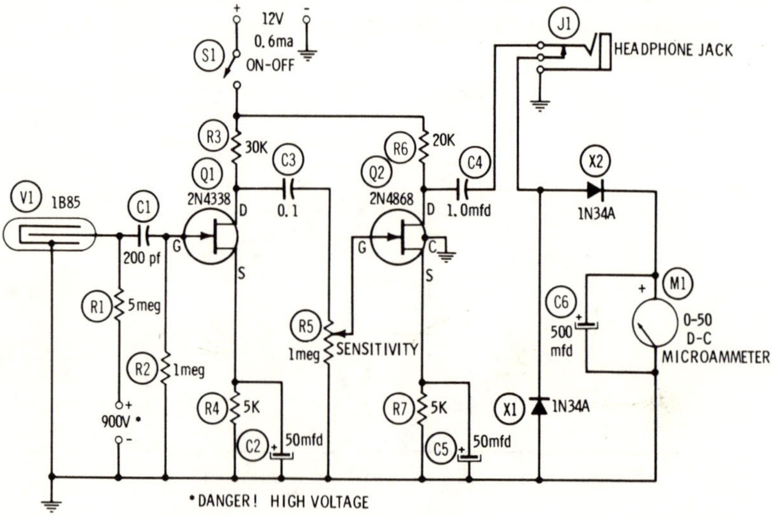

Fig. 7-20. Geiger counter.

A jack (J1) also is provided for alternate use of headphones; and since this is a closed-circuit jack, withdrawal of the phone plug automatically switches the meter back into the circuit. Current drain is 0.6 ma at 12 vdc.

The amplifier receives its input signal from a Victoreen 1B85 G-M tube (V1) biased by 900 vdc (Less than 10 ma). This is a beta-gamma particle counter. The type of self-contained supply selected for this high voltage will depend on the expense to which the builder can go. It can vary from three 300-volt photoflash batteries (Eveready No. 493, or equivalent) connected in series to one of the factory-built, miniature, dry-battery-operated high-voltage modules. Between these limits are numerous build-it-yourself units operated from flashlight cells (1.5 to 6 volts) of the transistor-oscillator, buzzer, and push-button–and–capacitor types. The ON-OFF switch in the high-voltage supply must be ganged with the instrument switch, S1.

All fixed resistors are 1/2 watt. Capacitor C1 is a 1000-volt ceramic unit; C3 and C4 are 100-volt plastic, and C2, C5, and C6 are 25-dcwv electrolytics. All wiring must be short and direct. The gate electrode is internally connected to the case of the 2N4338 (Q1), so this FET must be mounted clear of wiring, chassis, and other components. A fourth pigtail of the 2N4868 (Q2) is internally connected to the metal case of this FET and should be grounded as shown, for shielding.

The count-rate range of the instrument may be determined by exposing the 1B85 tube to various calibrated radioactive samples, noting the corresponding deflections of meter M1, and graduating the scale of the meter accordingly.

APPENDIX A

Terminal Guide for Diodes and Transistors

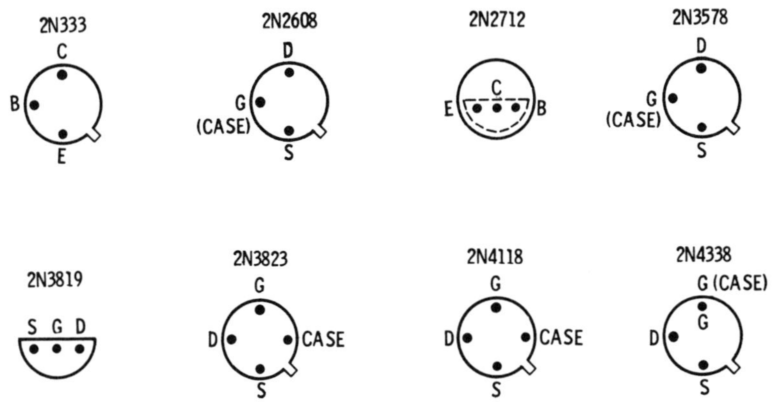

2N4340

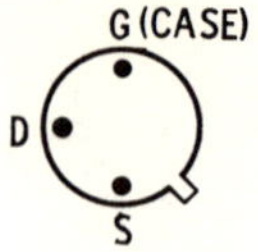

2N4868

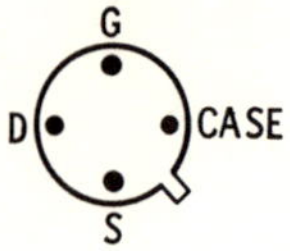

P-102

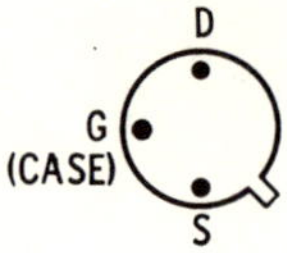

C632

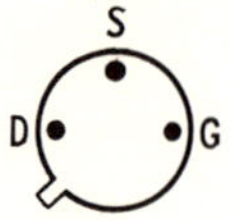

S1M

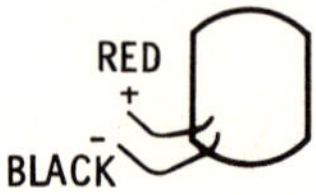

TIS14

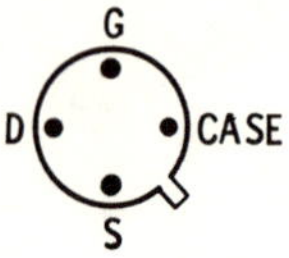

TIX33

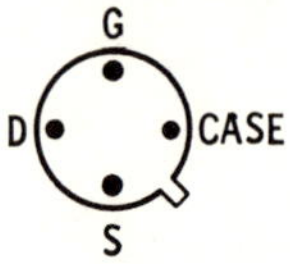

U183

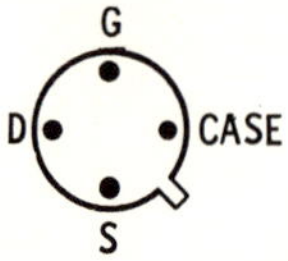

U222

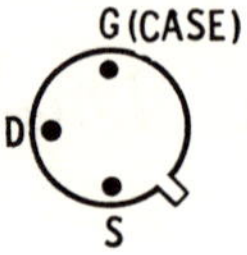

APPENDIX B

Manufacturers of Semiconductors

1N34A	GE, Sylvania	2N4118	Siliconix
1N295	GE, Sylvania	2N4338	Siliconix
1N3605	GE, Sylvania	2N4340	Siliconix
2N169A	GE	2N4868	Siliconix
2N333	GE, Raytheon, Texas Instruments	P-102	Siliconix
		C632	Crystalonics
2N2608	Siliconix	S1M	International Rectifier
2N2712	GE	TIS14	Texas Instruments
2N3578	Siliconix	TIX33	Texas Instruments
2N3819	Texas Instruments	U183	Siliconix
2N3823	Texas Instruments	U222	Siliconix

APPENDIX C

Directory of Manufacturers

AEROVOX. Aerovox Corporation. New Bedford, Massachusetts 02741.

ALLIED RADIO. Allied Radio Corporation. 100 N. Western Avenue, Chicago, Illinois 60680.

AMPEREX. Amperex Electronic Corporation. Providence Pike. Slatersville, Rhode Island 02876.

CLEVITE. Clevite Corporation, Piezoelectric Division. 232 Forbes Road. Bedford, Ohio 44014.

CRYSTALONICS. Crystalonics, Incorporated. 147 Sherman Street. Cambridge, Massachusetts 02740.

EVEREADY. Union Carbide Corporation, Electronics Division. 270 Park Avenue. New York, N. Y. 10017.

GE. General Electric Company, Semiconductor Products Department, Electronics Park, Syracuse, New York.

INTERNATIONAL RECTIFIER. International Rectifier Corporation. 233 Kansas Street, El Segundo, California 90245.

MILLEN. James Millen Manufacturing Co., Inc. 150 Exchange Street. Malden, Massachusetts 02148.

MILLER. J. W. Miller Company. 5917 So. Main Street. Los Angeles, California 90003.

RAYTHEON. Raytheon Co., Semiconductor Operation, 350 Ellis Street. Mountain View, California.

SIGMA. Sigma Instruments, Inc. 170 Pearl Street. Braintree, Massachusetts 02185.

SILICONIX. Siliconix Incorporated. 1140 W. Evelyn Avenue. Sunnyvale, California 94086.

SPRAGUE. Sprague Electric Company. 233 Marshall Street. North Adams, Massachusetts 01247.

SYLVANIA. Sylvania Electric Products, Inc. Semiconductor Division. 100 Sylvan Road. Woburn, Massachusetts 01801.

TEXAS. Texas Instruments, Incorporated. P.O. Box 5012. Dallas, Texas 75222.

TRIMM. Trimm, Incorporated. 400 Lake Street. Libertyville, Illinois 60048.

U. S. SONICS. U. S. Sonics Corporation. 63 Rogers Street. Cambridge, Massachusetts 02742.

UTC. United Transformer Corporation. 150 Varick Street. New York, N. Y. 10013.

VICTOREEN. Victoreen Instrument Division. 10101 Woodland Avenue. Cleveland, Ohio 44104.

Index

A

A-c a-f relay, 104-105
  r-f relay, 95-96
  voltmeter, electronic, 117-118, 119-120
Action, FET, 14
A-f amplifier
  bandpass, 30-34
  band-suppression, 34-35
  l-c–tuned, 30-31
  notch, 34-35
  r-c–coupled, 24-25
  r-c–tuned, 32-34
  transformer-coupled, 25-26
A-f milliammeter, 131-132
  oscillator, 56-57
    l-c–tuned, 45-47
    phase-shift, 47-48
    transformer-feedback, 44-45
    Wien-bridge, 48-50
  preamplifier, 20-22
  relay, 104-105
  signal tracer, 145-147
  source follower, 22-24
Agc amplifier, 38-39
All-wave receiver, 63-64
Amplified photocell, 97-98
Amplifier, a-f
  bandpass, 30-34
  r-c–coupled, 24-25
  transformer-coupled, 25-26
Amplifier
  audio agc, 38-39
  audio squelch, 71-72
  auxiliary, 37
  bandpass, 30-34
  band-suppression, 34-35
  bipolar-transistor, 26-27
  d-c, 39-41
  final, 78-79
  grounded-gate, 79-80
  headphone, 37-38
  i-f, 36-37
  modulated-light, 107
  push-pull, 78-79
  r-f, 79-80
  video, 36

Audio agc amplifier, 38-39
mixer, 30
oscillator, 127-128
squelch amplifier, 71-72
Audio-frequency meter, 124-126
Auxiliary amplifier, 37-38

B

Background, FET, 11-12
Balanced modulator, 84-85
Bandpass a-f amplifier, 30-34
Band-suppression a-f amplifier, 34-35
Beat-frequency oscillator (bfo), 66-67
Bipolar, definition of, 15
Bipolar-transistor amplifier, 26-27
Breakdown voltage, 17
Broadcast receiver, 59-61, 62

C

Capacitance, input, 17
reverse transfer, 17
Checker, capacitance, 135-137
inductance, 133-135
Code-practice oscillator, 57-58
Colpitts a-f oscillator, 56-57
Conductance, output, 17
Constant-current source, 103-104
Converter, crystal-controlled, 65-66
Counter, Geiger, 148-149
Crystal frequency standard, 121, 122-123
oscillator, 53-56
Crystal-controlled converter, 65-66
transmitter, 74-75, 76, 77-79
Current, dark, 98
Current source, constant, 103-104
Cutoff current, 17
C-w monitor, 86-87
transmitter, 74-75, 76, 77-79

D

Dark current, 98
D-c amplifier, 39-41
relay, 94
voltmeter, electronic, 112-113, 114-117
Depletion layer, 13
Detector, diode, 90-91
modulated-light, 107
Diode detector, 90-91
Dip oscillator, 129-131
Dissipation, 18
Distortion meter, harmonic, 141-145
Doubler, push-push, 81-82
Drain-coupled multivibrator, 50-51

E

Effect, field, 10-11
Electronic a-c voltmeter, 117-118, 119-120
d-c voltmeter, 112-113, 114-117
Exciter, four-band, 84

F

FET action, 14
background, 11-12
operation, 12-15
performance, 15-16
ratings, 17-18
structure, 12-15
terminal guide, 151
FET, symmetrical, 13
tetrode, 18
Field effect, 10-11
Field-effect transistor; see FET
Final amplifier, 78-79
Forward transconductance, 16, 17
Four-band exciter, 84
Frequency meter
audio, 124-126
heterdoyne, 139-141
Frequency
multiplier, 80-81
standard, crystal, 121-122

G

Geiger counter, 148-149
Generator, sound-marker, 126
Grounded-gate amplifier, 79-80
Guide, terminal, 151

H

Harmonic-distortion meter, 141-145
Headphone amplifier, 37-38
Heterodyne frequency meter, 139-141

I

I-f amplifier, 36-37
IGFET, 18
Impedance meter, 137-139
Inductance chart, 134
  checker, 133-135
Input capacitance, 17
Inverter, paraphase, 28-29
  phase, 27-28

L

Layer, depletion, 13
L-c–tuned oscillator, 45-47
Light meter, 147-148

M

Manufacturers, 153-156
Marker generator, sound, 126
Meter, audio-frequency, 124-126
  capacitance, 135-137
  distortion, 141-145
  frequency, heterodyne, 139-141
  harmonic distortion, 141-145
  impedance, 137-139
  light, 147-148
Milliammeter, a-f, 131-132
Millivoltmeter, a-c, 119-120
Mixer, audio, 30
Modulated-light amplifier, 107
  detector, 107
Modulating transmitters, 87-88
Modulator, balanced, 84-85
Monitor, c-w, 86-87
  phone, 90-91
MOS FET, 18
MOST, 18
Multiplier, frequency, 80-81
Mulitivibrator, drain-coupled, 50-51
  source-coupled, 51-52

N

NF; **see** noise figure
NFET, definition of, 13
Noise figure, 18
Notch a-f amplifier, 34-35

O

Operation, FET, 12-15
Oscillator, audio, 56-57, 127-128
  l-c tuned, 45-47
  phase-shift, 47-48
  transformer-feedback, 44-45
  Wien-bridge, 48-50
Oscillator
  beat-frequency, 66-67
  code-practice, 57-58
  Colpitts, 56-57
  crystal, 53-56
  dip, 129-131
  Pierce, 55-56
  r-f, 52-53
  variable-frequency, 127-128
Output conductance, 17

P

Paraphase phase inverter, 28-29
Performance, FET, 15-16
PFET, definition of, 13
Phase
  inverter, 27-28
    paraphase, 28-29
  shifter, 107-109
Phase-shift oscillator, 47-48
Phone monitor, 90-91
Photocell, amplified, 97-98
Photoelectric field-effect transistor; **see** photo-FET
Photo-FET, 98-99
  relay, 98-99
Pierce crystal oscillator, 55-56
Pinch-off voltage, 14-15, 17-18
Preamplifier, a-f, 20-22
Push-pull final amplifier, 78-79
Push-push doubler, 81-82

Q

Q-multiplier, 67-68

R

Ratings, FET, 17-18
R-c–controlled timer, 99-101
R-c–coupled a-f amplifier, 24-25
Receiver, all-wave, 63-64
  broadcast, 59-61, 62
  regenerative, 62, 63-64
  superheterodyne, 59-61
Regenerative receiver, 62, 63-64
Relay, a-c, 95-96
  a-c a-f, 104-105
  d-c, 94
  photo-FET, 98-99
  r-f, 95-96
  sound-operated, 105-107
  touch plate, 96-97
Resistor, FET as variable, 101-103
Reverse current, 17
  transfer capacitance, 17
R-f amplifier, 79-80
  oscillator, 52-53
  relay, 95-96
  signal tracer, 145-147

S

"Selectojet," 69-71
Shifter, phase, 107-109
Signal tracer, a-f r-f, 145-147
Single-control transmitter, 88-90
Sound-marker generator, 126
Sound-operated relay, 105-107
Source, constant-current, 103-104
Source follower, a-f, 22-24
Source-coupled multivibrator, 51-52
Squelch amplifier, 71-72
Standard, frequency, 121, 122-123
Structure, FET, 12-15
Superheterodyne receiver, 59-61
Symmetrical FET, 13

T

Terminal guide, FET, 151
Tetrode FET, 18
Theory, field-effect, 10-11
Timer, 99-101
"TNT" transmitter, 88-90
Touch-plate relay, 96-97
Tracer, signal, 145-147
Transconductance, forward, 16, 17
Transfer capacitance, reverse, 17
Transformer-coupled a-f amplifier, 25-26
Transformer-feedback oscillator, 44-45
Transmitter, crystal-controlled, 74-75, 76, 77-79
  single-control, 88-90
  "TNT," 88-90
Transmitters, modulating, 87-88

U

Unipolar, definition of, 15

V

Variable phase shifter, 107-109
  resistor, FET as, 101-103
Variable-frequency oscillator, 127-128
Video amplifier, 36
Voltage, pinch-off, 14-15
Voltage-variable resistor, 101-103
Voltmeter, electronic, a-c, 117-118, 119-120
  d-c, 112-113, 114-117

W

Wien-bridge oscillator, 48-50